CIM Interfaces

KU-131-737

672825

CIM Interfaces

Concepts, standards and problems of interfaces in Computer Integrated Manufacturing

B. Scholz-Reiter
Institute of Systems Analysis, Berlin University of Technology, Berlin, Germany

Translated by Christoph Törring

Technical advice by Eoin Molloy
University College, Galway

CHAPMAN & HALL
London · Glasgow · New York · Tokyo · Melbourne · Madras

Published by Chapman & Hall, 2–6 Boundary Row, London SE1 8HN

Chapman & Hall, 2–6 Boundary Row, London SE1 8HN, UK

Blackie Academic & Professional, Wester Cleddens Road, Bishopbriggs, Glasgow G64 2NZ, UK

Van Nostrand Reinhold Inc., 115 5th Avenue, New York NY10003, USA

Chapman & Hall Japan, Thomson Publishing Japan, Hirakawacho Nemoto Building, 6F, 1–7–11 Hirakawa-cho, Chiyoda-ku, Tokyo 102, Japan

Chapman & Hall Australia, Thomas Nelson Australia, 102 Dodds Street, South Melbourne, Victoria 3205, Australia

Chapman & Hall India, R. Seshadri, 32 Second Main Road, CIT East, Madras 600 035, India

First English language edition 1992

© 1992 R. Oldenbourg Verlag GmbH, München

Original German language edition – CIM Schnittstellen – 1991, R. Oldenbourg Verlag GmbH, München

Printed in Great Britain by St Edmundsbury Press Ltd, Bury St Edmunds, Suffolk

ISBN 0 412 45160 3 0 442 31626 7 (USA)

Apart from any fair dealing for the purposes of research or private study, or criticism or review, as permitted under the UK Copyright Designs and Patents Act, 1988, this publication may not be reproduced, stored, or transmitted, in any form or by any means, without the prior permission in writing of the publishers, or in the case of reprographic reproduction only in accordance with the terms of the licences issued by the Copyright Licensing Agency in the UK, or in accordance with the terms of licences issued by the appropriate Reproduction Rights Organization outside the UK. Enquiries concerning reproduction outside the terms stated here should be sent to the publishers at the London address printed on this page.

The publisher makes no representation, express or implied, with regard to the accuracy of the information contained in this book and cannot accept any legal responsibility or liability for any errors or omissions that may be made.

A catalogue record for this book is available from the British Library

Library of Congress Cataloging-in-Publication data available

D
621.78
SCH

Contents

Introduction

In the modern industrial production environment, the number of computer applications is steadily increasing. Over the past three decades the potential for automation, especially in actual manufacturing, has been put into practice by applying ever more sophisticated computer aided manufacturing tools and methods.

To support the planning and control of material requirements, capacities, and production deadlines, from the 1960s onwards electronic data processing applications have been introduced into industrial manufacturing enterprises. And since the beginning of the 1980s, computer aided systems have also been increasingly employed in the areas linked to manufacturing, i.e. especially in the fields of development and design, as well as process planning.

With an increasing saturation of companies with the mentioned systems, even a continuous upgrading of the systems keeping step with the rapidly advancing state of the art of information technologies will achieve only a marginal growth in efficiency. Further benefits of rationalization are almost impossible to achieve. Where data processing applications have been vigorously introduced, the organization and sequence structures in the above-mentioned functions and fields (development, design, and process planning) have already reached a near optimum. It is possible to attain almost optimal solutions in certain areas, although, with regard to the enterprise as a whole, they do not necessarily coincide with the overall optimal solution.

In order reach such an overall optimum, a comprehensive global perspective covering all areas related to the execution of orders, and thus also the data processing applications on which they rely, is necessary. This leads us to the conclusion that the various individual areas concerned with order execution in our present enterprises are often too strictly segregated. This division of labour frequently has the effect that basic data must be generated again and again in the various areas. If this is aggravated by a lack of synchronization, batch processing is the result. Both factors lead to unnecessarily high costs and long lead times.

The main requirement today is not only increased efficiency, but also a high degree of flexibility in order to ensure the profitability of an enterprise. Seen from a classical point of view, this would result in a dilemma with regard to rationalization. Extra capacities have to be retained in order to be able to manufacture customer oriented products in many variants with shorter lead times (Hirsch-Kreinsen 1986).

A decrease in the division of labour among the above-mentioned areas, made possible by organizational measures which utilize modern information and communication technologies, may provide a way out of this dilemma, or may at least lead to its minimization. Information is an important factor in manufacturing and must be recognized as such. The future potential for rationalization thus will not lie in the installation of isolated computerized solutions in the different areas concerned with order execution (island solutions), but rather in their coupling and therefore in the utilization of synergical effects. This fundamental concept is often referred to by the term computer integrated manufacturing (CIM) which was coined by Harrington in the United States as early as 1973 (Harrington 1973). Harrington established the basic principle of the production cycle, which starts with the development of a product, continues through manufacturing and shipment to the customer, and ends with service and maintenance.

Computer integrated manufacturing in a more concrete sense includes the coupling of all areas linked to the actual manufacturing operations, whereby data processing techniques on the one hand support the geometrical/technological domains (computer aided design (CAD), computer aided process planning (CAPP), computer aided manufacturing (CAM), computer aided quality assurance (CAQ)), and on the other hand the organizational/administrative domains (mainly production planning and control systems (PPC)). In a wider sense it also embraces the common areas of business administration such as sales, financial administration, and personnel management, whereby specialized programming systems or common tools such as the computer aided office (CAO) environment play a role.

The above-mentioned data processing tools show different degrees of development as individual solutions. Integrated systems are already being applied mainly in the organizational/administrative area, and in the geometrical/technological field the coupling of CAD and NC is relatively far advanced. The main task for the future lies in achieving a coupling of the geometrical/technological and the organizational/administrative areas, and also in finding ways of connecting the various CA systems used in the geometrical/technological area with each other. Also, integration on a higher level with the coupling of several companies, the implementation of which until now has been attempted only by the automobile industry and their parts suppliers, will gain in momentum.

Coupling does not only require the solution of technical problems, but in the first place necessitates administrative and organizational measures. The qualification level of the staff plays a major role. 'Integration' without the corresponding organizational consequences leads only to marginal benefits. It is not possible simply to adapt the currently used means and methods. Before attempting integration on a data processing level, organizational integration with regard to products, processes, methods, etc., has to be accomplished (for the important area of the planning of CIM information and communication systems see for instance Scholz-Reiter 1990).

This book intends to provide a first overview of the general possibilities of the coupling of CIM applications and the related problems as an aid for those who wish to study them or put them into practice. In this book the focus is put primarily on interfaces at the application level. Man-machine interfaces or interfaces such as V.24 or X.25 will therefore be disregarded. The scope of the book is largely limited to the problems involved in the coupling of computer aided systems commonly applied in medium-sized mechanical engineering companies. In Germany, this type of company constitutes an important proportion of the total number of manufacturing enterprises.

Chapter 1

Fundamentals of computer integrated manufacturing

1.1 Components of computer integrated manufacturing

The concept of computer integrated manufacturing is based on the principle of the production cycle developed by Harrington. The production cycle starts with the development of a product, then moves on to manufacturing and delivery to the customer and ends with service and maintenance. As within the operational sequence organization these functions are closely interconnected, it is not expedient to regard the single functions as isolated from each other. According to Harrington, in the final state the production cycle can be described as a sequence of data processing operations (cf. Harrington 1973).

Extending Harrington's concept, the components involved in computer integrated manufacturing (CIM) therefore are the data processing tools underlying the production cycle.

Much has been written with regard to the definition and the substance of CIM without a comprehensive consensus having been found as yet. The CIM definitions offered by manufacturers, mostly from the information technology branch, differ widely. They mostly vary in the range of what they consider as part of CIM. (For manufacturer's definitions by Control Data, DEC, Hewlett-Packard, IBM, Philips, Siemens, Sperry, Data General, ICL, Mannesmann Kienzle, Prime, Wang and Nixdorf see for instance Scholz 1985 and Geitner 1991.)

In the present context, the interpretation of the term CIM as it has been proposed by several neutral international organizations, independent of company affiliations, will be discussed; this interpretation has won the approval of users, manufacturers, consultants, etc. A clarification of terms and a definition of the components which can be attributed to computer integrated manufacturing will be attempted by referring to the publications of the AWF (AWF 1985) and the CASA/SME (cf. Appleton 1986). Among others, a clarification of terms and references was also achieved within the framework of the ESPRIT project 'Design Rules for A CIM System' (Yeomans *et al.* 1985).

The CIM definition by CASA/SME

The Computer and Automated Systems Association (CASA) of the Society of Manufacturing Engineers (SME) of the United States, also known as the organizer of the CIM-Fair and the Autofact Congress, can be compared with the Gesellschaft Produktionstechnik (ADB - Society of Production Technology) of the Verein Deutscher Ingenieure (VDI - Association of German Engineers) in Germany. As early as 1980, the CASA/SME published a presentation of computer integrated manufacturing in order to provide a common set of terms for its members. In Germany, a similar step was taken five years later by the AWF. The graphical representation of CIM originally published by CASA/SME is shown in Fig. 1.1. The ring surrounding the wheel represents various influencing factors (man and his degree of expertise as the human factor, productivity as the economic factor and computer technology as the technological factor) for the development of CIM. The wheel itself contains four functions, namely engineering design, manufacturing planning, production control and factory automation. If the individual functions are connected with each other and operate with a common database, an integrated system architecture is created which is represented by the hub of the wheel.

The original CIM picture of the CASA/SME has undergone a number of changes in recent years (Fig. 1.2) (cf. Appleton 1986). This development has resulted in the realization that CIM, apart from factory automation and functions indirectly related to the operational performance such as design (product/process) and production planning and control, is also linked to common business administrative tasks such as manufacturing management, strategic planning, finance, marketing and human resource management.

A further innovation was the addition of information resource management and communications between the different functions. A common database alone therefore is insufficient for achieving integration. The all-embracing nature of the CIM wheel reflects the idea promoted by CASA/SME that CIM has to be viewed as a concept embracing the company as a whole.

In the following sections, the CIM wheel is further discussed, taking into account the current state-of-the-art in data processing applications within the individual functions.

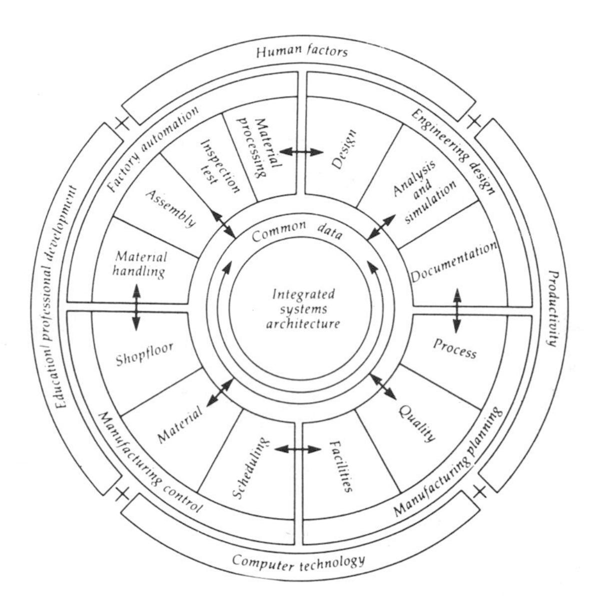

Fig. 1.1 CIM wheel - CASA/SME 1980.

The outer ring

The common business administrative tasks related to CIM are located on the outer ring of the wheel. They mainly form the connection of the company to the outside world in general. Data processing applications can be found in the most diverse areas. Most software systems applied in these areas were originally self-styled developments, which are increasingly being replaced with commercial standard software packages. Currently this software is installed primarily on mainframes. Overlaps of its functionality exist mainly with the software of the production planning and control.

The inner ring

On the inner ring of the wheel, the functions closely related to the operational performance of the company are situated. Data processing applications of the development and design area are computer aided design (CAD), simulations, analysis programs such as the finite element method (FEM) as well as drawing storage and management, for instance with the help of group technology (GT).

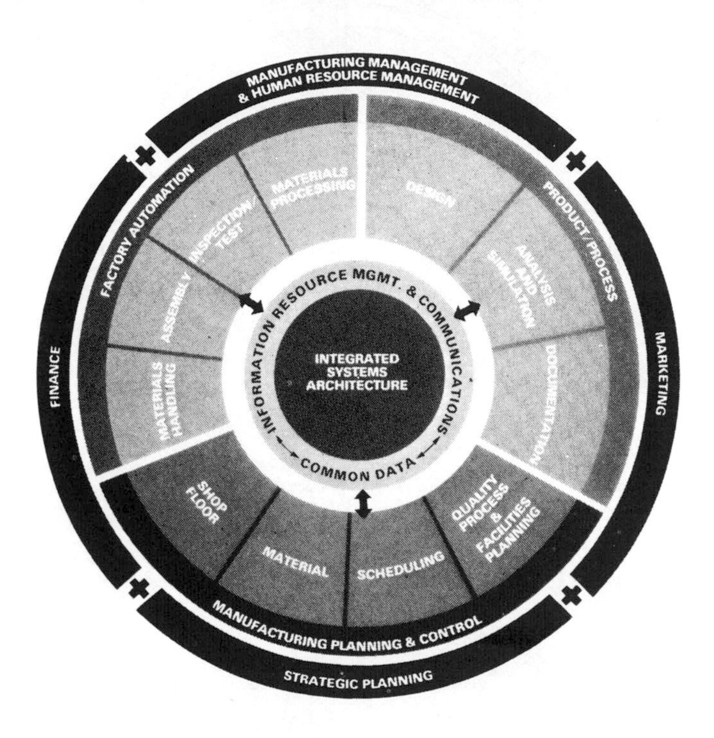

Fig. 1.2 CIM wheel - CASA/SME 1986 (Appleton 1986).

The types of data found in this area are diverse: drawings, technical specifications, and bills of material. In manufacturing companies, the data itself are often in considerable disorder. Frequently there are several types of part numbers, more than one group technology system, many kinds of bills of material, a number of different CAD systems each having its own sort of computer internal representation of geometric data, etc. The applied software rarely runs on the same hardware, resulting also in a large number of different hardware systems.

The second group of applications on the inner ring of the wheel is attributed to process planning and production planning and control. It comprises tasks such as routing generation, resource planning, material requirements planning, capacity planning, order distribution and supervision, but also the planning of quality assurance (quality process and resource planning). In the USA, software in the production planning and control area mostly runs on IBM mainframes although the software itself is more often than not supplied by sundry software houses and not by IBM itself. As in the common business administrative area, the software packages - which at least are integrated within themselves - have a modular structure and their single components can also be bought and applied. Therefore a company rarely has purchased and installed all modules of such a package, which in turn frequently results in functional overlaps and data redundancy (one example: material requirements planning and purchasing systems).

The third group on the outer ring includes the automation of the manufacturing installations. Examples are robots, numerically controlled machines, flexible manufacturing systems and computer aided measuring and testing methods. This area is characterized by the extreme heterogeneity of the systems involved, the diversity of which being much more pronounced than in the previously mentioned groups of functions.

The hub

The statements made above have already shown that at present within these groups of applications there are serious impediments with regard to integration. Wider still is the gap between the domains on the inner ring of the CIM wheel. There are few suppliers covering all three sectors. Therefore little or nothing has been done by the suppliers with regard to interfaces, not to mention the integration of the various groups of applications. Information and communication management, represented by the hub of the wheel which links everything, is intended to serve as the information management and communication control function between the single areas. It operates on a common, integrated database.

The CIM definition by the AWF (cf. AWF 1985)

The Ausschuss fuer wirtschaftliche Fertigung e.V, (AWF - Committee for Economical Manufacturing) includes in its definition of CIM in the broadest sense all data processing systems in all company areas related to manufacturing which are to be integrated. CIM encompasses the interaction by information technology of CAD, CAPP, CAM, CAQ, and PPC systems (Fig. 1.3). Its aim is to achieve an integration of the technical and the administrative functions involved in manufacturing. This implies the use of a common database which integrates the different areas (AWF 1985). In accordance with the recommendations laid down by the AWF, the above mentioned abbreviations are briefly explained in the following paragraphs.

In the area of pre-production functions, on the one hand CAD systems (computer aided design) and on the other hand CAPP systems (computer aided process planning) are applied.

CAD systems directly or indirectly support the development and design effort. In a more narrow sense, digital object representations of parts, assemblies, circuit boards, etc., are generated and manipulated with graphical interaction. In a wider sense, common technical calculations with and without graphical input and output are included. The finite element method (FEM) for the calculation of thermal or mechanical stresses is one example.

CAPP systems are applied in the area of process planning. They use the results generated in the design process. CAPP systems cover data processing systems which support the planning of the operation processes and the operation process sequence as well as the selection of methods and resources necessary for the manufacturing of the objects, and - very importantly - they generate the control programs for the resources (NC programming).

Quality assurance still forms a part of the pre-production functions but also overlaps into the area of production functions. CAQ (computer aided quality assurance) indicates the support by data processing tools of the planning and execution of quality assurance, which on the one hand includes the generation of test schemes, test programs and control values and on the other hand the execution of measuring and test methods.

The application of CAM systems (computer aided manufacturing) relates to the area of production executing functions. Their tasks are the technical control and supervision of the resources during the manufacturing of the objects, i.e. the manufacturing, handling, transport and storage. As diverse as these tasks are the components applied to fulfil them. They can consist of single CNC machines (computerized numerical control), several numerically controlled manufacturing installations which are controlled by a host computer (DNC systems - distributed numerical control), industrial robots (IR), automated material handling systems (AMH), electronically controlled high-rise storage systems, etc. When some of these components grow together, flexible manufacturing cells (FMC) can be formed around one processing machine or, if several processing machines are involved, flexible manufacturing systems (FMS) can emerge.

PPC systems (production planning and control systems) are applied as higher level instruments for the organizational planning, control and supervision of the production processes with regard to volumes, delivery dates and capacities. A differentiation can be made between the main functions of production program planning, material requirements planning, capacity planning on the planning side and scheduling, manufacturing order release and manufacturing order execution supervision on the control side.

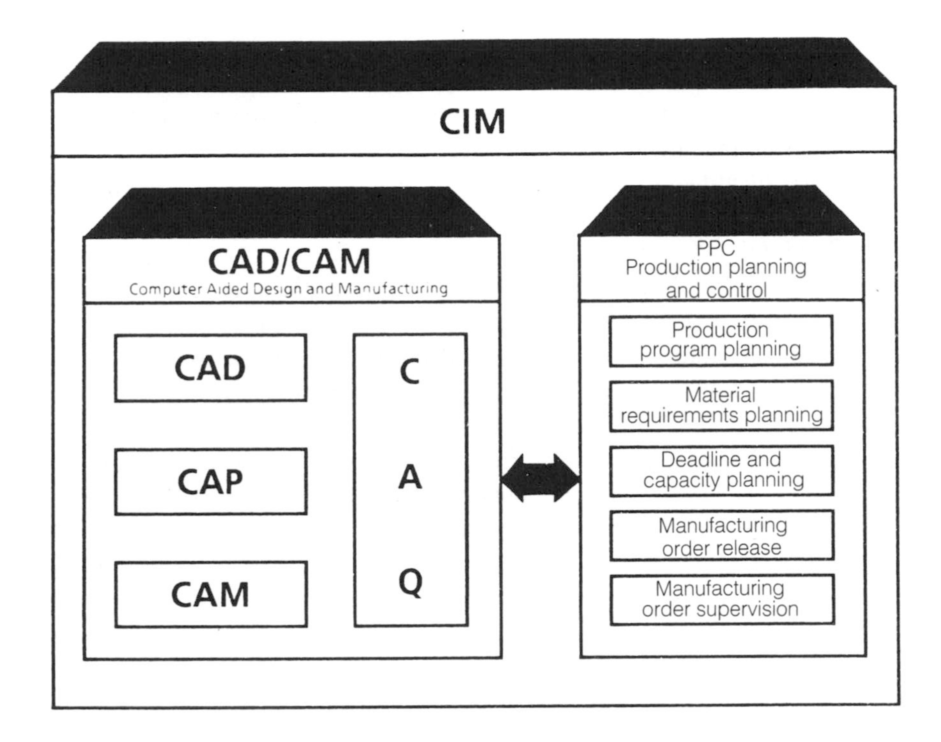

Fig. 1.3 Definition of CIM terms (AWF 1985).

Manufacturing data collection (MDC) is not regarded as a CIM component in its own right. Rather, MDC is viewed as part of the functions which plan, control, and supervise production. MDC is therefore included in the CIM functions PPC, CAQ and CAM.

Information flows in computer integrated manufacturing

The main information flows involved in computer integrated manufacturing were clearly outlined by Helberg. CAD generates product models and product describing data which are transformed by CAPP into routings and control programs for the CAM systems. The PPC systems generate and manage all operational data which are used for controlling in the CAM area. CAQ on a short-term basis corrects deviations in the manufacturing process and in the long run influences the development of products and methods with regard to quality assurance (cf. Helberg 1987).

Helberg's outline does not include further necessary or desirable informational connections between the systems, such as a connection of CAD/CAPP and PPC for an accompanying calculation during design and routing generation, or feedback from manufacturing to planning. Furthermore, at least in the case of single-parts manufacturing, processes like design and process planning can be regarded as elements of the lead time of an order and therefore can be planned and controlled by the PPC system in the same way as the actual manufacturing and assembly processes. In that case a corresponding feedback becomes necessary. Fig. 1.4, closely following Helberg's model, therefore gives an outline of the components together with the additional informational relationships existing between them. It becomes evident that computer integrated manufacturing is a system of intermeshed control loops.

Computer integrated manufacturing necessitates additional communication technologies indispensable for the connection, ranging from the passing-on of data carriers over a direct coupling of components to the application of networks. Also a common logical database which can be represented physically by central or distributed databases is an element essential to the implementation of the concept. It will form the basis for a comprehensive information and communication management system.

1.2 Current state of information processing in manufacturing

The current state of computer supported applications within manufacturing companies is normally characterized by isolated solutions in the individual application areas. In many companies, some areas of production do not have any computer support at all.
Especially in small and medium-sized production, a comprehensive connection of the process functions in areas related to the manufacturing process is rarely found. This statement is founded on several independently conducted empirical surveys in which manufacturing companies featuring an advanced application of computers were examined or CIM experts were interviewed.

In the following, some of the results of these studies (Förster et al. 1985, Helberg *et al.* 1985, Köhl *et al.* 1988, Köhl *et al.* 1989, Schultz-Wild *et al.* 1989), which cover the situation in the former Federal Republic of Germany until the year 1989, will be summarized and evaluated with an eye on the prevailing development trends.

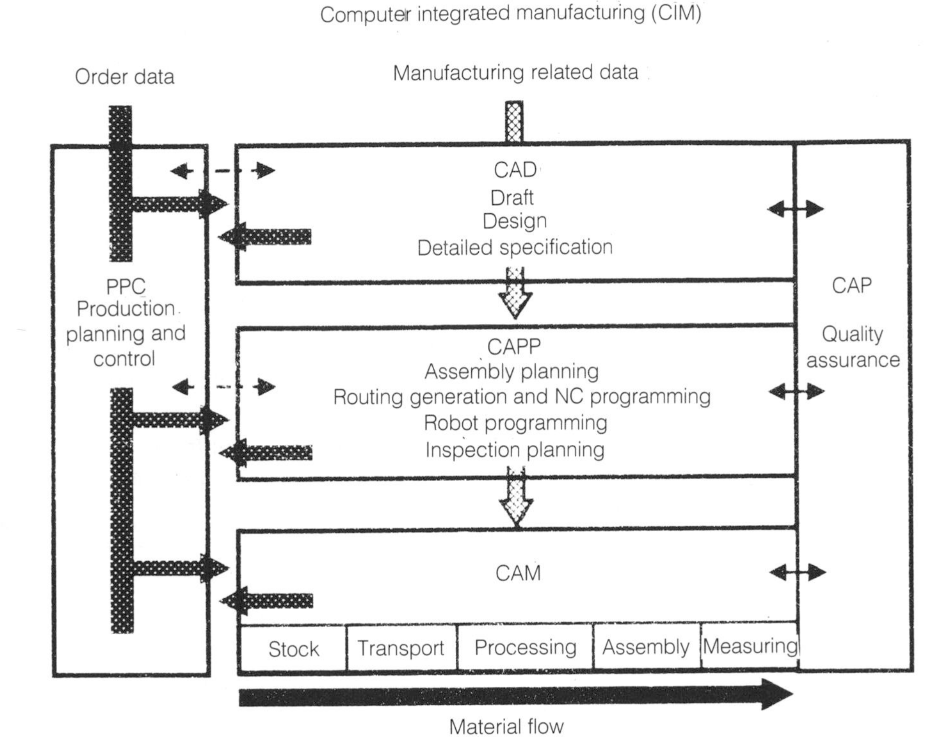

Fig. 1.4 Functions and information flows in computer integrated manufacturing (with reference to Helberg 1987).

The bigger the company, the higher the number of functions in the manufacturing area which are supported by data processing tools. This implies that more small and medium-sized companies than big ones are currently planning to introduce computers in such areas. In all companies, computer support in the production planning and control area is most advanced. Roughly a quarter of the companies have some experience with PPC systems. The inquiries also show, though, that only a few companies have implemented integrated PPC systems with a full functional range. The main emphases - 60-70% - are placed on basic data management and material requirements planning. Above all in the production program planning and the capacity planning area computer support is limited. Its share amounts to only 40% of companies with PPC experience. Still smaller is the share in the control area. In a mere one third of the companies with PPC experience, order release and supervision and MDC are computer supported.

Another essential factor for the evaluation is the state of the art of the currently installed PPC systems. Most of them are batch oriented programs based on data management systems. Standardized database systems for all applications in the PPC area are rarely used. The share of dialogue processing and of user controlled inquiry functions in the installed PPC systems is on the average merely 30%. Consequently, batch processing is still wide spread.

The share of CAD applications in manufacturing companies is roughly 20%, more than half of which feature dialogue processing and user controlled inquiry functions. The range of functions implemented in the applied systems is an open question. As a rule, they will tend to be systems used merely for creating design drawings rather than, for instance, geometric modellers.

With routing generation and NC programming applied in more than 50% of the companies, the CAPP area is relatively far advanced. No statements can be made about the quality and the state of the art of the applied systems.

In the CAM area, the application of computers for machine control is relatively advanced with a share of 35%. Handling and transport systems amount to roughly 5%. Computer support of quality assurance can be found in about one tenth of the companies. Again, no statements can be made about the state of the art of the applied systems.

It can be stated with regard to systems applied in areas related to manufacturing that about the same number of companies are planning to introduce such systems by the early 1990s. Therefore the number of companies applying these systems will have doubled within five years. The introduction of computer controlled manufacturing machines and installations in the immediate manufacturing area will show only a marginal growth, the current level of application being already comparatively high. Growth rates here are about 100% for material flow systems, assembly systems, and the DNC sector.

The surveys show that companies as a rule are disinclined to replace existing systems with new ones offering more functions. This explains the huge gap in the development stage especially between PPC systems available on the market and those which are actually applied. Nearly 80% of all companies, however, are enhancing the functions of their PPC systems. Normally they aim at a functional enhancement in individual areas, whereby 'classical' functions such as bill of material management, purchasing and the processing of customer orders are also addressed. The introduction of computer aided systems is not mentioned as the primary reason for this enhancement.

The trends mentioned above do not permit any statement about the integration of the various systems.

In 1986/87, less than 10% of the companies had achieved at least one sort of system integration (cf. for the following Schultz-Wild *et al.* 1989). However, about 15% were planning to implement a system integration by the early 1990s. Fig. 1.5 gives a good impression of the current state of integration of the systems applied for the different functions.

The figure shows that in 1986/87 the number of system integrations put into practice was still very low, but that there were plans to quadruple their number by the beginning of the 1990s. But even then the level of integration would still be extremely low. Furthermore it has to be taken into account that from the numbers alone no conclusion whatsoever can be drawn with regard to the extent of the integration. The integration actually achieved may be on a low level, or it may not even cover the whole necessary information exchange between the two systems involved.

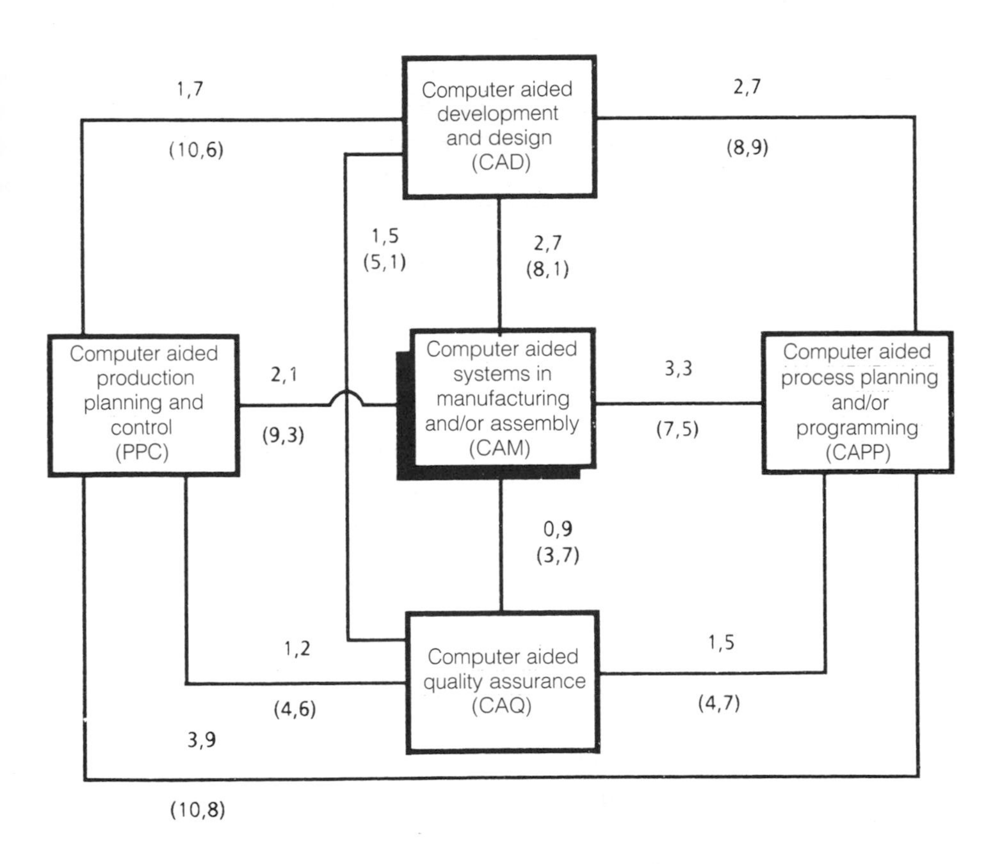

Fig. 1.5 System integration within companies - implemented in 1986/87 or planned (manufacturers of capital goods, whole number: 1096, data as a percentage) (Schultz-Wild *et al.* 1989).

Two main trends of integration can be identified: on the one hand an integration of the geometrical/technological systems CAD-CAPP-CAM, and on the other hand the integration of the administrative PPC systems and CA applications. However, integration attempts of the second type are still mainly in the planning phase.

In spite of the previously mentioned disadvantages of the currently applied systems, it is expected that PPC systems will form the kernel of computer aided manufacturing systems in the future, as computer support in production planning and control, in contrast to other areas, is already relatively wide spread. Another kernel will be formed by CAD systems which are also already relatively wide spread and in practice are technologically more advanced than the PPC systems.

The high economic risks involved, however, form a serious obstacle to a further advance of the integration process within companies. Medium-sized companies often find it very difficult to meet the high costs arising from integration on a larger scale. The problem of proving the economical benefits already arises with the introduction of isolated solutions. It grows even more difficult once synergic effects are taken into account.

In the framework of this overview it was not possible to enlarge upon the spread of information technology with regard to the types of companies which actually employ it. More comprehensive and detailed information is given on all the above by Schultz-Wild *et al.* 1989.

1.3 Model of a sequence structure

The continuity of computer supported process links is of prime importance for the implementation of computer aided manufacturing.

In the following, the process of carrying out an order by a manufacturing company is considered under the aspect of data continuity. Problems of interfacing which inevitably occur when connecting current hardware and software systems are not taken into account. On the one hand, the logical data flow from sales to design, routing generation, NC programming and robot programming is studied. On the other hand, the data flow from the technical to the administrative area - the PPC system - is examined.

The aim is to give an example of the whole process, before the problems of interfacing are addressed in detail in the subsequent chapters (cf. for instance Seifert 1986).

A customer's order is accepted by the sales department and input into a computer-based order management system. If a standardized product is requested, the order can be handed over directly to the production planning and control system as a primary demand. Otherwise the order has to be processed by the design department with the aid of a CAD system. According to whether the requested article is new, a modified one or a variant, earlier identical or similar solutions are examined. The design engineer, in a graphically interactive dialogue with the CAD system, generates designs which are then stored in a design database which is part of the CAD system. They can be accessed by calculation programs such as, for instance, FEM modules with which the designed components can be examined with regard to mechanical or thermal stresses. Depending on the results of these calculations, the engineer modifies the geometry of the component or possibly conducts a repeated calculation.

The whole design process normally results in hardcopy drawings and bills of material which are usually handed over to process planning manually.

The primary task of computer aided process planning (CAPP) is the generation of the routing for the manufacturing process. It contains mainly the operation sequences necessary for manufacturing, together with the corresponding utilities and tools, the techological process data, the set-up times and the advance times. These data depend on the geometrical and master information laid down by design. When computer aided process planning tools are applied, an already existing routing is examined in which, for a group of similar parts, the technological sequence of all operations involved is laid down in a network structure. According to certain selection criteria, each operation is examined with regard to whether it can be skipped or must be included in the operation sequence. Now for each operation which has been included, the other data as mentioned above are determined. At present, automatic routing generation is only possible in certain cases, i.e for parts featuring rotational symmetry. For parts with complex shapes automatic routing generation is extremely complicated. When processing standardized routings of which only the parameters have to be modified or when planning the adaptation of already existing routings, strong computer support is available. Often, though, in the area of routing generation computers are only applied as a means for obtaining information, storing data or for use as word processors.

As the routing data strongly depend on the design of the part, a connection with the design database is advisable.

Further tasks of the process planning sector involve completing the master information of the part laid down by design and transforming the design bill of material to a production bill of material, i.e. by checking the availability of the components.

To make the routing complete, the planning area of the quality assurance (CAQ) generates test schemes which are mainly based on the generated CAD data.

NC programming is a further task of process planning. The necessary data can be derived from the geometry of the part generated by design and partially from data contained in the routing. In interaction with the process planner, the NC module generates so-called part programs which in a second step are transformed into machine control data. Postprocessors adapt the results of the NC module which are normed as CLDATA to the actual machine controller. Considerably more complex than NC programming is robot programming. Apart from the geometry of the part, the geometry of the handling robot has to be taken into account for generating collision-free motion paths. Preferably, these motion paths are interactively simulated on a graphical screen.

All results of process planning are stored in a database and are then placed at the disposal of the manufacturing area.

Having by way of example shown the data flow within the geometrical/technological area of the order execution process, in the following paragraphs the process within the administrative area, the PPC system, is outlined. Relying mainly on the design and routing data, the PPC system organizes and controls the actual manufacturing process.

To begin with, for planning a production program the following questions have to be answered for a given period of time taking into account predicted or already accepted orders:

- Which products are to be manufactured?
- In what volumes are these products to be manufactured?
- What is the deadline for the manufacture of these products?

Here the available production capacity has to be taken into account. Proceeding from the results of the production program planning and depending on the current actual amount of orders, on the one hand systematic material requirements planning and on the other capacity planning are conducted. Based on the manufacturing bills of material generated by process planning, the material requirements planning lays down quantities and deadlines for the single parts and assemblies which are derived by taking into account the lead times of lower level production steps.

During the material requirements planning step, the calculation of the quantities of all parts on all production levels, i.e the quantity aspect, is the dominant one; the time aspect is only taken into account with regard to lead times. The task of capacity planning, however, is to generate a precise timeframe for the manufacturing process taking into account the available capacities. This task proceeds from the routings generated by the process planning area, data about primary demands, and data about resources. The results are manufacturing orders with a fixed timeframe which can be executed with the available capacities; when due, they are released for production (CAM) after a short-term verification of the availability of materials, utilities, etc., has been conducted. The current state of the production progress in the actual manufacturing process is checked by the order supervision function. This ensures a reaction in case of a deviation from the planned data. The gathering of the data required by order supervision is performed by the manufacturing data collection function (MDC), which registers the current data and hands them over to order supervision.

In a similar way, data for computer aided quality assurance are gathered and are compared with the reference values of the test schemes, etc. The feedbacks have to be processed by the order supervision for the planning functions. All functions of the production planning and control system are integrated and operate in a common PPC database.

The process in the PPC area described above relates mainly to the manufacturing of standardized products. It is relatively simple as previously generated bills of material and routings are ready at hand.

If, however, products not yet or not completely designed and planned products are to be manufactured, production planning has to start at a much earlier stage, even before design and process planning have begun. The creative and planning steps should also be regarded as an element of the lead time of an order. In spite of the many difficulties, the PPC system should lay down and control deadlines and capacities for them also.

Having presented a schematic model of the sequence in Fig. 1.6, in Fig. 1.7 the data processing applications concerning the route of an order in the tool construction area of a big industrial company are outlined as an example. The tool construction can be regarded as an order-oriented single parts or batch manufacturing area. The integration in this case is not ideal. There is no data link, for instance, between the CAD and the PPC system.

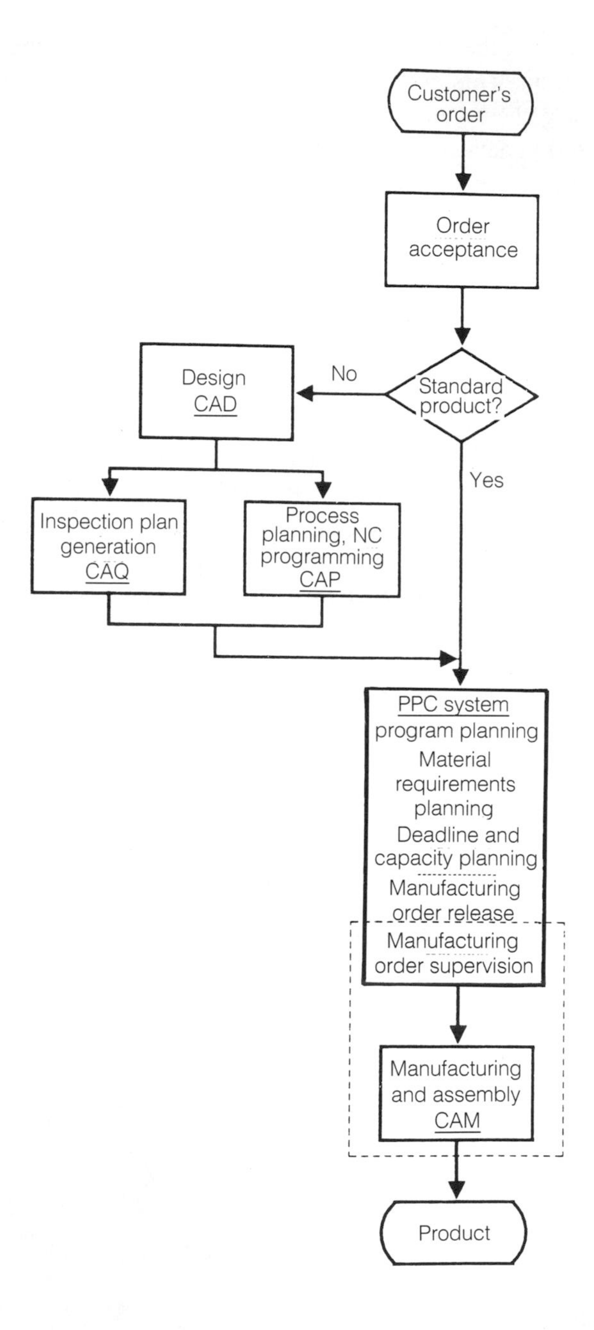

Fig. 1.6 Simplified flow chart of the route of an order from order procurement to the finished product.

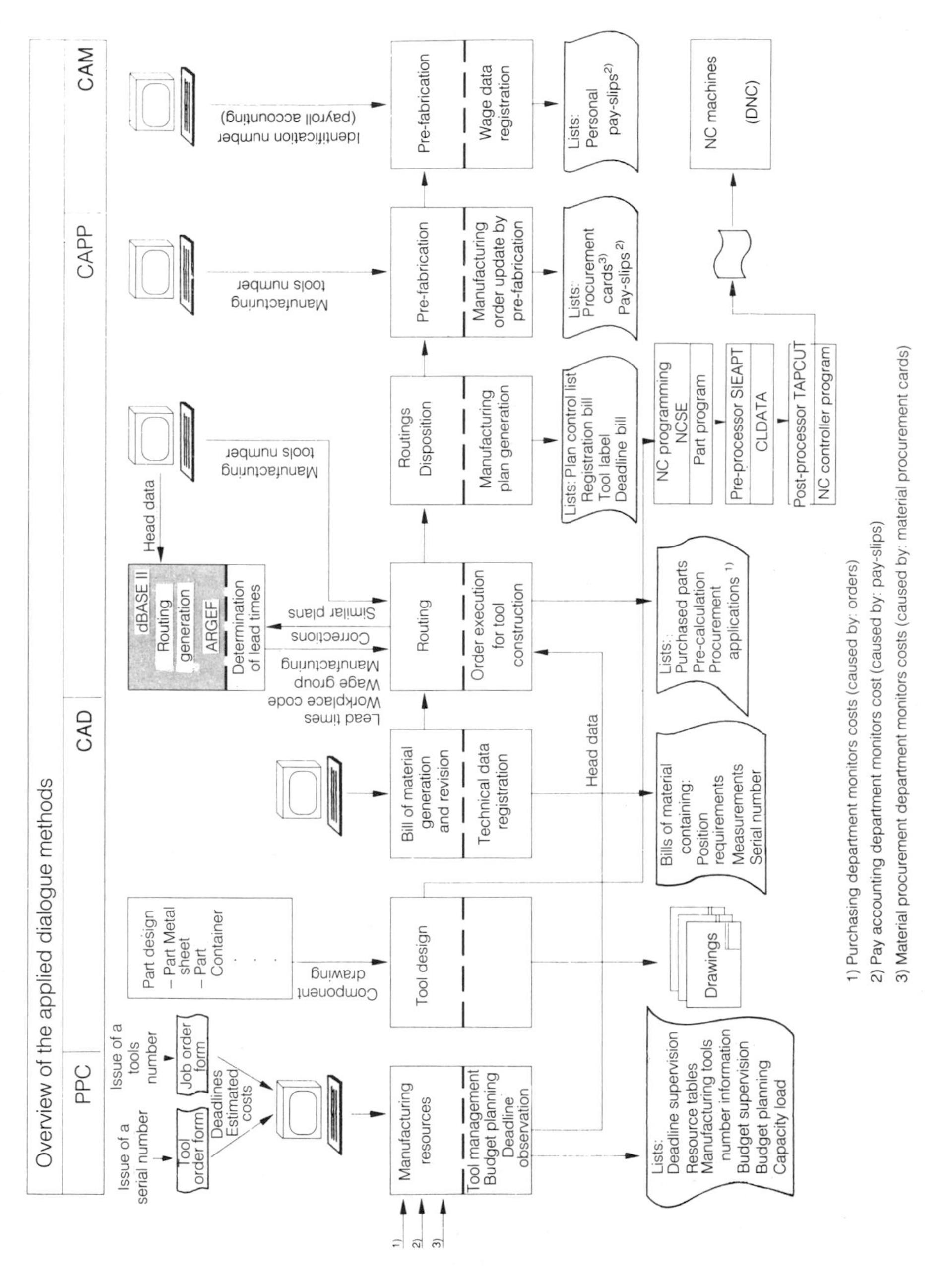

Fig. 1.7 Data processing applications involved in the route of an order in the tool construction area of an industrial company (Scholz *et al.* 1987a).

1.4 Interface - integration - coupling - connection

The main idea of computer integrated manufacturing is the 'integration' of the data processing applications in the manufacturing area, which as a consequence of Taylorism (Taylor 1911) were mostly designed and established as isolated solutions. Driven forward by the rapid advance of data and information processing technologies, the main objective is to attain a continuation of process chains without breaks caused by the media involved (cf. Scheer 1988). Process chains constitute an organizational/technical attribution of tasks and/or operations to a certain process. By forming continuous process chains, functions which today are as a rule isolated from each other - as for instance design and NC programming - are organizationally, as well as technically, connected with each other. The aim therefore does not lie in the optimization of the single functions but rather of several functions linked together as one process. The functions interact by information exchange. In the following chapters, such an information exchange forming the connection of two functions will be understood as an interface (cf. Scholz-Reiter 1991). Items such as the man-machine interface or interfaces such as V.24 or X.25 are therefore not discussed here. Deterministic quantities of an interface are the distribution of the functions to the various applications and the ensuing information exchange. Common syntactics and semantics of the applications which are to be connected enable the information exchange to take place. The physical implementation with the help of networks, databases and hardware interfaces is generally not considered. (For types and requirement criteria of interfaces see, for instance, Brenig 1990.) With regard to the ISO/OSI reference model (cf. Chapter 4), most of the interfaces discussed here are those of the seventh layer, the application layer. As a consequence of the heterogeneity of applications, a standardization in this layer is difficult to achieve. However, a classification of applications and their properties can lead to a realization of their common characteristics and their similar communication behaviour and therefore can open possibilities for standardization. Examples are interfaces of product definition data (cf. Chapter 3) or MMS (manufacturing message specification) with the specific companion-standards in the framework of MAP (manufacturing automation protocols) (cf. Chapter 4) designed for communication with devices on the manufacturing level, i.e. CNC machines, robots, etc. In this context, integration means the reduction or even the complete avoidance of interfaces, whereby the latter possibility in the future will also be confined to pure scientific speculation due to restrictions of the capacity and complexity of the applications.

Integration (for the functions of integration see, for instance, Heilmann 1989) in this sense has a twofold meaning. According to Scheer (Scheer 1988) it can be either data integration or operations integration (see Fig. 1.8). Operations integration means the recombining of formerly divided applications. Lead-in times and data transfer times are reduced. The division of labour which previously often was meaningful can in many cases be reversed because the application of new technologies improves the formerly limited human capacity for information processing. Therefore tasks and processes can be reintegrated into one application or reallocated to one operator's position. Interfaces as they were defined above are avoided by operations integration. Data integration continues the division of labour between the various applications, but avoids a functionally divided storage of data. The whole process chain is - at least theoretically - based on a unified, common database. Data transfer times are thus avoided. In practice such systems can be found, for instance, in the production planning and control area.

In the context of CIM, the term integration with reference to data integration is often applied to mere coupling which, regarded under the aspect of data processing technology, is a low-level solution.

However, the coupling of components involved in computer integrated manufacturing means merely a connection by data technology of two independent program systems. The program systems normally have different and independent data and storage structures. Two systems are coupled by an intermediate data file and/or a coupling procedure. This implies the problem of data redundancy. The number of coupling procedures can be very high, as specially adapted coupling procedures have to exist between each pair of application programs.

Because of the computerized data processing, however, even coupling can result in higher efficiency and data security.

Integration differs from mere coupling in the sense that the overall system, i.e. all applications, is based on a unified model with a common data and storage structure valid for all applications: coupling procedures are superfluous; the problem of data redundancy is avoided; and the applications communicate with each other. The preconditions for such an integrated system are outlined in more detail below.

Hirsch-Kreinsen in this context has pointed out a very important aspect (cf. Hirsch-Kreinsen 1986): integration may not merely be regarded from the viewpoint of information technology as the application of software packages, hardware components and the implementation of data and information transfer. Integration is linked to a high degree to process organization. The sequence of processes as a whole is systematized, tightened, and made continuous. As a result the division of labour can be partially removed, the remaining areas can be better synchronized, and performing the same work twice can be avoided.

The term 'connection' as it has been used on the previous pages and will be used in the following chapters can mean both coupling and integration.

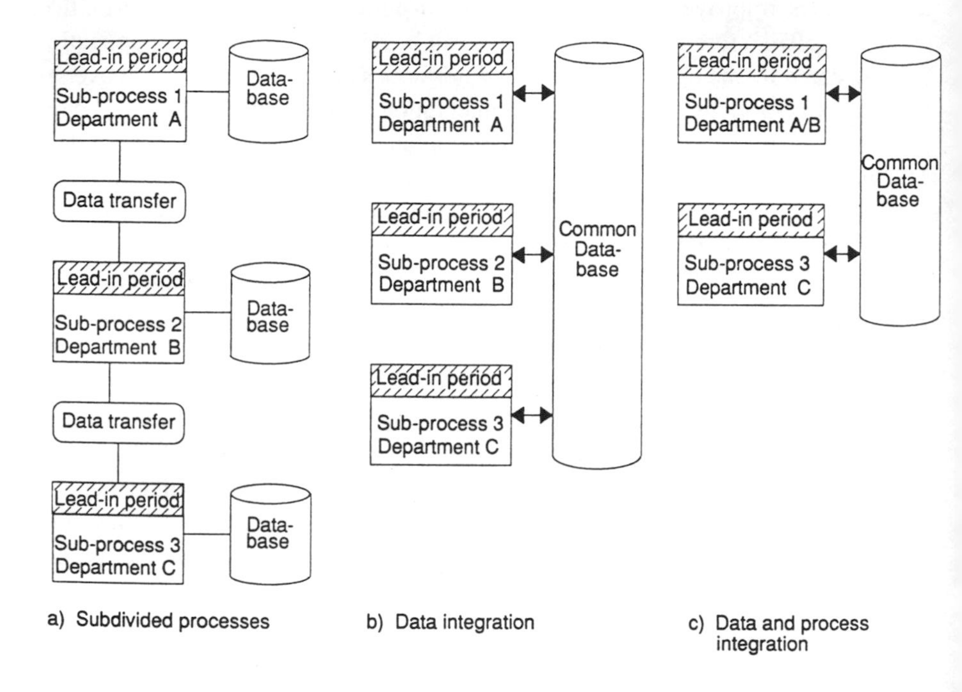

Fig. 1.8 Reintegration of functionally divided operations (Scheer 1988).

1.5 Summary

At the moment, no comprehensive consensus regarding the definition of CIM exists. In the present context, the components of computer integrated manufacturing are meant to be the following data processing tools available in the production cycle:

- PPC systems;
- CAD systems;
- CAPP systems;

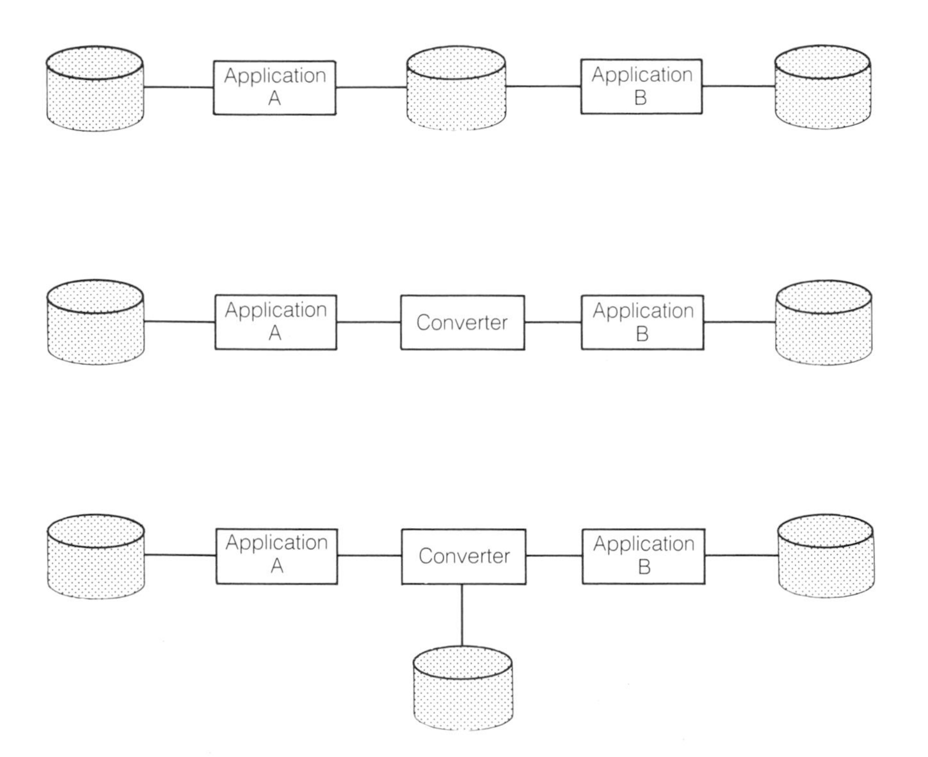

Fig. 1.9 Principal means of coupling by data technology of two applications A and B.

- CAQ systems;
- CAM systems.

The continuity of computer supported process chains is of prime importance for the implementation of computer integrated manufacturing. Process planning and control serves as a high-level instrument for the organizational planning, control and supervision of the production processes with regard to aspects of volume and capacity. For the time being, the interfaces between the components form a major obstacle to the implementation of computer integrated manufacturing.

Two emphases of connection can be detected:

- the connection of CA systems of the geometrical/technological area with each other, and
- the connection of the more administrative functions with the geometrical/technological functions.

In contrast to the coupling of components which implies only the data technological connection of two different program systems, data integration requires a unified model on which all applications are based. Integration may not be regarded merely under the aspect of information technology, but is strongly linked to process organization.

At present, a connection of process functions can be found only in certain areas of companies. The focus is mainly put on the integration of PPC functions with each other and on the coupling of CAD and NC programming.

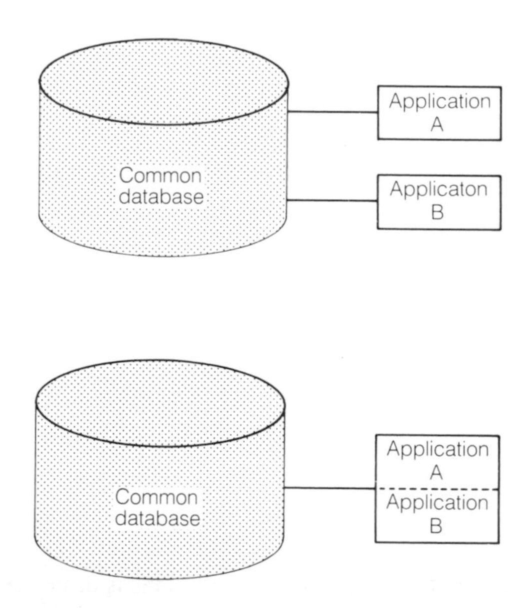

Fig. 1.10 Means of integration of applications.

Chapter 2

Basic means of connecting data processing applications

In this chapter the different basic means of connecting data processing applications will be discussed. Concepts for all-round connection which will be essential for an explicit integration in the factory of the future, will be addressed in Chapter 7.

2.1 Linkage by human interaction

One linking element between two data processing applications often found in the real company environment is man. This is *a priori* not a technical connection by data processing, but an organizational linkage of two systems not connected by any data processing means.

A number of different types of this human interaction do exist. As an example, two non-interactive applications are linked by an operator who selects data from the output lists of one application and then feeds them as input data to a second one, or simultaneously and at the same place works with two different interactive applications and transfers information between both. In this way the applications may run on two independent computers and the dialogue may be performed with the help of two terminals. The applications may also run on the same terminal using window technology. The human interaction can consist of a selective conversion or an interpretation of the data. As this form of linkage is not a connection by data processing means, it will not be discussed in further detail here.

Fig. 2.1 Human interaction between two applications A and B.

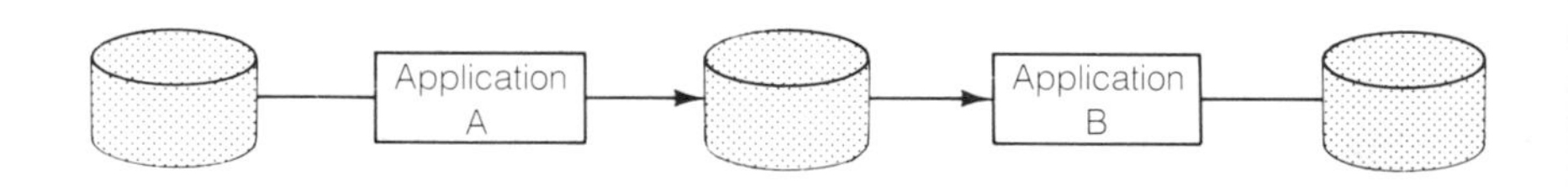

Fig. 2.2 Coupling by formatted file.

2.2 Coupling by formatted file

Coupling by a formatted file presents a relatively simple form of a data processing connection between two applications.

As a rule, the data and storage structures of both applications differ. Application A stores data in a file with a predetermined format. This format has to be known to application B to enable it to read the file and process the data it contains. This type of coupling exists only for one pair of applications and data transfer is uni-directional. Often difficulties occur when the programs are altered. Normally this sort of coupling is a user-programmed solution. The data exchange on the hardware side can be done off-line by a number of means, such as punched paper tape, magnetic tape, diskettes, etc., but also on-line solutions by linking up two computers via point-to-point connection or local area networks (LAN) are possible (cf. Chapter 7).

The main advantage of this type of coupling of two applications lies in the fact that data already available to one system do not have to be transferred manually to a second one for further processing. The work-load is thereby reduced and transfer errors can be avoided.

2.3 Coupling by converter programs

This solution of the coupling of two applications features a converter program which adapts the data structure of application A to the one used by application B. Data transferred from application A to application B can be supplemented either by further stored data or by interaction between converter and user.

Converters between two defined application systems which sometimes can also handle a data transfer in two directions (A to B, B to A) are offered by software suppliers in growing numbers. Examples are CAD/PPC processors such as the system CADMIP by IBM which connects the PPC system COPICS with the CAD system CADAM and vice versa, or CAD/NC processors such as the system CADIS-NCS by Siemens which performs the coupling of the CAD system CADIS with the NC processor SIEAPT. This type of connection is mostly applied when both applications have different data structures, or when they even run on different computers and at different times independent of each other.

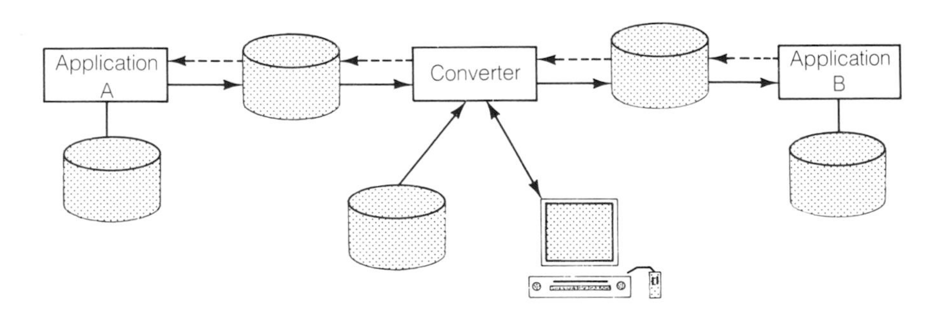

Fig. 2.3 Means of coupling by converter.

The fact that the converter is usually specifically adapted to connecting two applications poses a disadvantage. It means that for an all-round data transfer within an overall system with n applications a large number of converters (n x $[n-1]$) is necessary. Attempts to achieve standardization in order to reduce this problem are discussed in the further chapters.

The large number of converters can be reduced to $2n$ for all $n>3$ through standardization.

A dynamic information access of one application to another is not possible on-line.

The application of mailbox systems is a further development of data transfer by formatted file or converter.

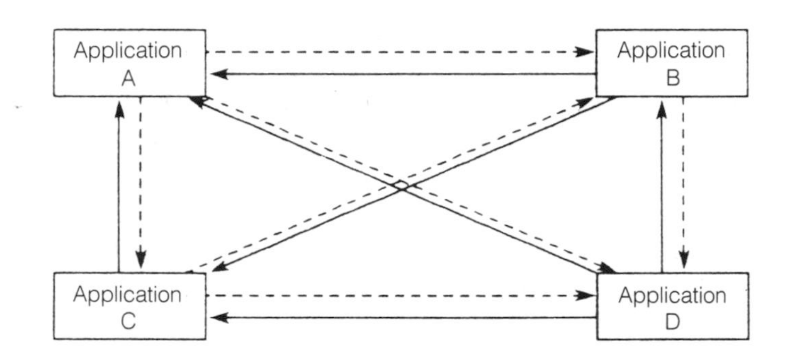

Fig. 2.4 Coupling without standardization (4 x 3 = 12 converters).

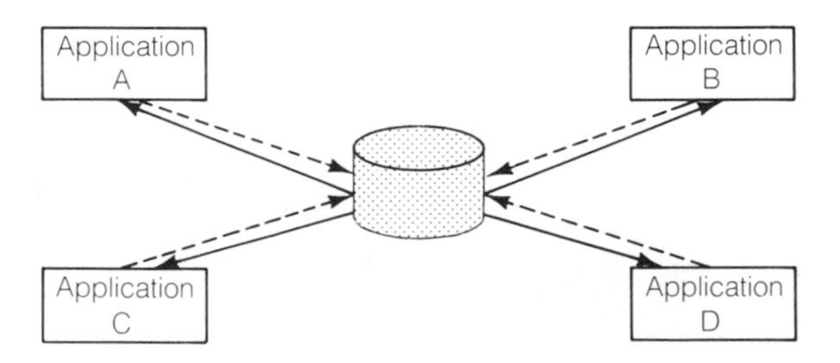

Fig. 2.5 Coupling with standardization (8 converters).

2.4 Connection via databases

The exchange of data via a database system enables the connection of numerous applications without involving a high number of intermediate files or converters. Basically, several possible combinations do exist.

One possibility is that the autonomous applications have their own data files and transfer to a joint database only those data which are necessary for communication. Another one is that the whole stock of data can be accessed by all applications via a central, standardized database system.

The concept of the first possibility entails the same problems of data redundancy as the previously mentioned types of connection.

In database systems, the data of several or all applications are gathered in the database. By means of a database management system which manages their access to the data stock, the applications have a logical view of the data. Thus they are independent of the storage structure of the data. One advantage database systems have over data file concepts lies in the possibility of multi-user operation whereby the integrity of the data is maintained in case of concurrent access. The database system checks the data integrity with the help of consistency conditions defined by the user, above all when operations resulting in enlargement, alteration or deletion of data are carried out. Data protection in case of a system breakdown, faults of data carriers or interruption of transactions are performed independently by the system. The multitude, heterogeneity and complexity of applications in computer integrated manufacturing pose a challenge when determining a suitable structure for the database systems. Should standardized database models for CIM be hierarchical, relational or have a network structure, or should special new concepts be developed?

2.4.1 Standardized database systems

For the last ten years, standardized databases have been used for administrative data processing applications. In computer integrated manufacturing they have been applied in the production planning and control area. Three different models can be identified: the hierarchical model, the network model and the relational model.

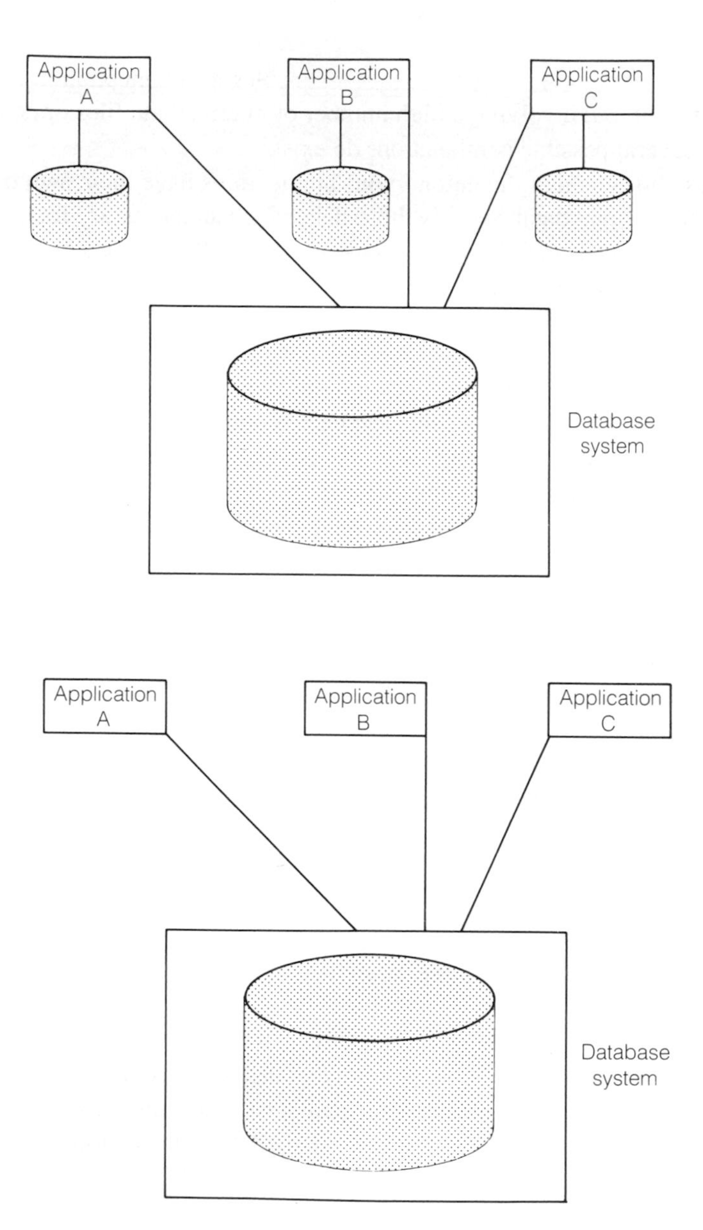

Fig. 2.6 Data communication via databases.

2.4.1.1 Hierarchical model

In database systems which are based on the hierarchical model, the correlations within the model which represents reality are laid down in a strictly hierarchical manner by using tree structures. The structure therefore is defined without ambiguity. As an example, the structural bill of material of a product P1 is represented by a tree.

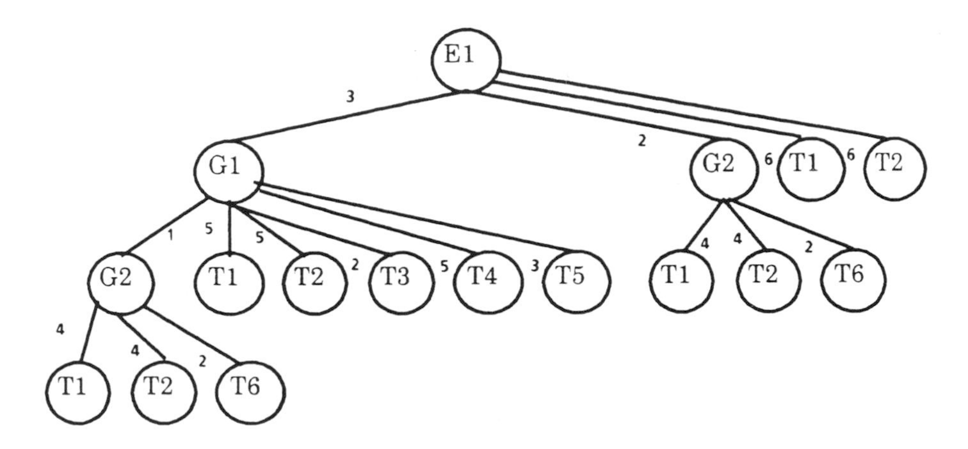

Fig. 2.7 Generative part structure of product P1.

The number of nodes (E1, G1..., T6) corresponds to the number of real entities (part trees of the final product, assemblies, components). The relationships between the entities (here analytically: consists of, synthetically: is used by) are represented by the arcs.

Seen from the viewpoint of data processing techniques, this data organization is efficient as it definitely determines the access paths. This representation, however, is not very user-supportive and complicates alterations, as the access to all entities has to move along the predetermined tree structure.

The hierarchical model also restricts the representation of n:m relationships. The reason for this restriction is that ambiguous relationships cannot directly be represented by hierarchical data models; they have to be split up into a 1:n and a 1:m relationship which together then represent a n:m relationship.

Due to these disadvantages, a hierarchical database structure has turned out to be impractical for administrative data processing applications.

Table 2.1 Structural bill of material

Step	Part number	Number of parts required
1	G1	3
2	G2	3
3	T1	12
3	T2	12
3	T6	6
2	T1	15
2	T2	15
2	T3	6
2	T4	15
2	T5	9
1	G2	2
2	T1	8
2	T2	8
2	T6	4
1	T1	6
1	T2	6

2.4.1.2 Network model

Here, the model symbolizing reality is represented by networks.

As in the hierarchical model, the entities are represented by nodes. However, the network model, in contrast to the hierarchical model, allows a direct representation of ambiguous relationships by a node having a number of input and output arrows. These directional arcs therefore have to be specified in order to identify their meaning.

In the network model, data connections and access paths have to be precisely determined, i.e. pre-programmed. The effort involved in generating such a data-

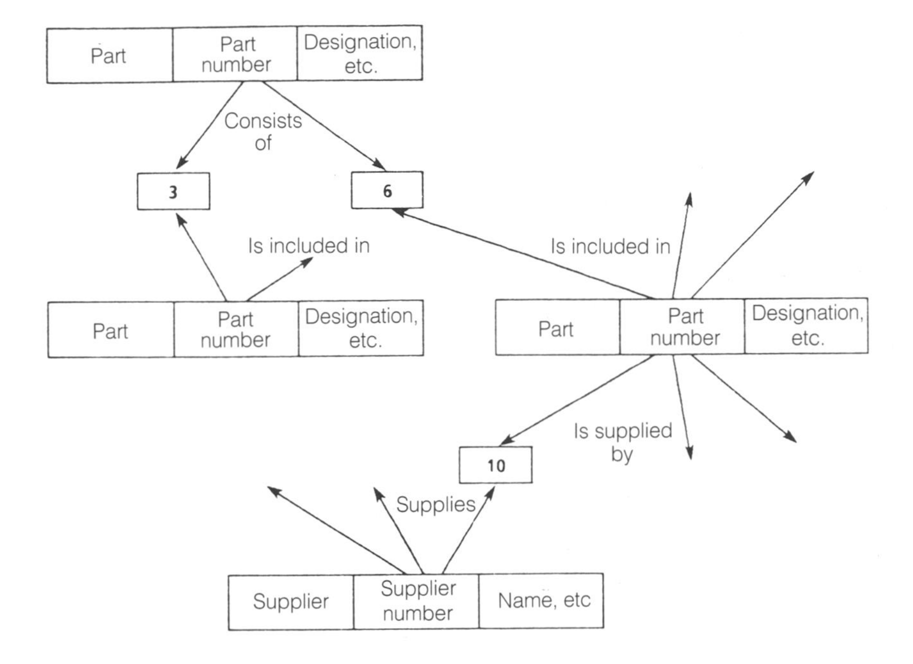

Fig. 2.8 Network mode: relationship between generative part structure and suppliers.

base is quite high, but the work-load put on the computer for its application is relatively low.

If within a company the tasks to be fulfilled are clearly defined, not undergoing many alterations, and many transactions have to be performed, as for instance in the PPC area, the application of databases organized according to the network model is useful.

2.4.1.3 Relational model

The relational model excels by its simplicity and clarity. In contrast to both of the models previously discussed, it does not distinguish between entities and relationships. The model is based on the relation, i.e. the two-dimensional table. Entities and relationships are both represented by tables (see F

The main advantage offered by databases organized accor model is that the data connection and the determination of acce to be defined *a priori*, but have to be determined only when an access to the database is actually taking place. The effort involved in creating such a database therefore is much lower than in the case of the network model, and alterations can be performed easily.

The fact that when a high data volume and complex transactions are involved the access times are relatively long constitutes a disadvantage. Relational databases are mainly applied in areas where tasks frequently change and/or experiences are yet few and therefore major alterations during the operation are to be expected.

For relational databases, a standardized database language, Standard Query Language (SQL) (ANSI 1986a, b) has been developed, which in a static form is included in application programs and in a dynamic form is used for interactive *ad-hoc* queries.

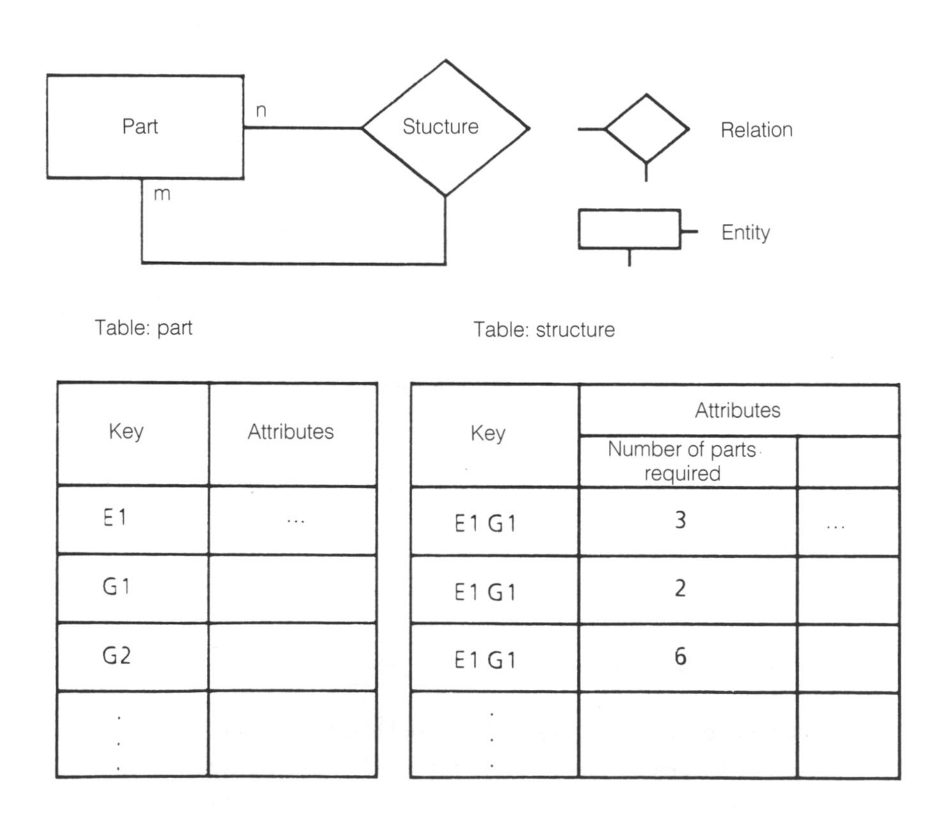

Table: part

Key	Attributes
E1	...
G1	
G2	
⋮	

Table: structure

Key	Attributes	
	Number of parts required	
E1 G1	3	...
E1 G1	2	
E1 G1	6	
⋮		

Fig. 2.9 Simplified relational model: generative part structure.

2.4.2 Non-standard databases

While standardized database systems can meet the demands of a PPC system, they appear not to be applicable in the CA area due to different requirements. Here so-called non-standard database systems are applied. Table 2.2 briefly explains the different demands on database systems by applications of the administrative area on the one hand and the geometrical/technological area on the other hand (cf. Küspert 1986)

A strong effort has yet to be undertaken for the further development of non-standard databases. Although during recent years much research has been conducted in this area, applicable solutions have yet to be found.

Two approaches can be identified. On the one hand, standardized database systems are enlarged by modules adapted to specific computer aided systems, and on the other hand special new data models for specific applications are developed. However, there are serious doubts about whether it is really possible to integrate

Table 2.2 Comparison of requirements on databases by administrative versus geometrical/technological applications (cf. Küspert 1986)

Administrative applications	Geometrical/technologically oriented applications
Numerous small objects	Big objects often too long for conventional databases
Simply structured objects (formatted data)	Quite complex structure of inhomogeneous data (unformatted data)
Relatively simple integrity conditions	Non-trivial integrity conditions
Relatively short transactions with access to few objects	Very long transactions (for instance CAD constructions)
Only current status is represented by the database	Former states or different versions or variants should be available (CAD), history management
Application of conventional non-intelligent VDUs	Workstation concept commonly applied
No real-time requirements	Real-time requirement (CAM, controllers, NC programs)

the isolated solutions currently prevailing in the technical area by the use of databases at all (cf. Wedekind 1986).

To achieve the representation of structures and behaviours of complex heterogeneous objects as they are commonly found in the area of product data, and in order not to leave the generation of structures solely to the applications, as it is normally the case with standardized databases, object orientation has become a focal point of the development (cf. Dittrich 1988, Nittel 1989). The database systems therefore support the definition of individual (abstract) data types and operations running with them. A further development effort aims at fulfilling the real-time conditions required above all by the production control area (cf. Wedekind 1987). Object oriented data buffers (caches) located in the main memory are one of the current approaches.

In the following, some examples are given for non-standard database systems in the protoype or project stage:

PHIDAS	was developed by the Philips research laboratory in Hamburg; it forms the integrating kernel of PHILIKON, a graphically interactive CAD system with a modular structure (Fischer 1982). It follows the CODASYL network model.
TORNADO	is a database system for CAD/CAM systems following the network model (Meen *et al.* 1982)
AIM-P	(Advanced Information Management-Prototype) is a project of the IBM Scientific Center in Heidelberg (Dadam *et al.* 1989).
R2D2	(Relational Robotic Database System with extendable data types) as a basis for an integrated robot programming and simulation system, R2D2 is based on AIM-P as a joint project of the University of Karlsruhe and IBM (Dadam *et al.* 1989).
DASDBS	(Darmstadt Database System) is being developed at the Technical University Darmstadt (Paul *et al.* 1987).

2.4.3 Central databases

Central databases form an ideal concept for integration. All applications have access to the central database and therefore all use uniformly up-to-date data.

However, this solution, being the highest level of integration, currently remains an illusion. A central database for applications in computer integrated manufacturing does not exist and is not expected to be developed in the foreseeable future. The reason for this can be found in the above mentioned demands placed on databases which - for the time being at least - cannot be met.

The power of today's computer systems, reasons of data security as well as the demand for constant availability of this vital nerve of a company are also arguments against a centralization of data in the manufacturing area. The enormous effort and the high costs faced by any company trying to simultaneously re-arrange the giant volume of data in the manufacturing sector will almost certainly render the acceptance of this solution impossible, even if it was feasible from a technical point of view.

The concept of a central database is therefore not discussed in further detail.

A better approach is to start from a common logical database of all applications which are based on distributed database systems.

2.4.4 Distributed databases

A common database is to be understood as a volume of data which are logically related. If this volume of data is physically distributed on database systems running on several computers located at different sites, they are called distributed databases. Distributed database systems offer several advantages:

- The different applications of computer integrated manufacturing also differ in their demands placed on the database systems. In the administrative area, standardized database systems are applied, while CA applications use non-standard database systems.
- The administrative area is normally supported by central all-purpose computers. The geometrical/technological sector usually relies on specialized personal computers and intelligent workstations due to the varying computing power and storage facilities required by the different applications. These computers are connected, for instance, by a local area network.
- When data transfer costs are relatively high, the overall costs can be reduced by storing and mainly using data at the place where they are generated.
- Local data management can take over responsibility for the accuracy and the security of the data.
- The predominantly local users enjoy faster access and inquiry possibilities when the data are stored locally.
- Data security is enhanced by physical distribution of the data volume.
- There is a higher degree of availability in case of errors.
- Distributed databases better match the frequently distributed organizational structure of enterprises.

A distributed database system consists of a number of local database systems, each having its own database management system and its own database, which communicate via a network for

- the transfer of inquiries;
- the transfer of resulting data;
- the management of distributed transactions;
- global inquiry optimization; and
- dynamic structure management (cf. Effelsberg 1987).

Thus, the network connects the single database management systems with each other. The application programs operate directly with their local database management systems. The system is responsible for the correct processing of inquiries to the whole database of the network.

The problem of heterogeneous, distributed database systems has not yet been solved. The communication protocols used by distributed databases are not standardized. As in the case of non-standard database systems, a considerable development effort has yet to be undertaken. Lately, interesting new approaches have been proposed which are known as the server-workstation concept (cf. Deppisch *et al.* 1986).

Specialized workstations are connected to a central computer (possibly a special database machine). The central computer serves as a database system for the workstations. It also functions as a server. The workstations also store local data.

A good example of this concept is an interactive CAD system where the processing of images which requires a relatively high number of calculations and is time-critical for the user, and whereby large volumes of data have to be transferred to and from the main memory, is taken over by a workstation. Thereby the central computer, functioning as a multi-user system, is relieved of these tasks. First, the required data are transferred from the server to the workstation (known as extraction, check-out). The involved objects are then excluded from further transactions. The data in effect are 'privatized'. After finishing the design process, the data are handed back to the server (known as injection, check-in) and again put at the disposal of other users. This concept is useful when, during the time of the transaction, the data are required by only one user. In the design process, this is normally the case.

A special variant is the concept of a federal database server developed by the Eidgenössische Technische Hochschule Zurich, Switzerland. The users store their private data on a complete local database system running on their workstation. The public data required by all or several users are stored centrally by the server.

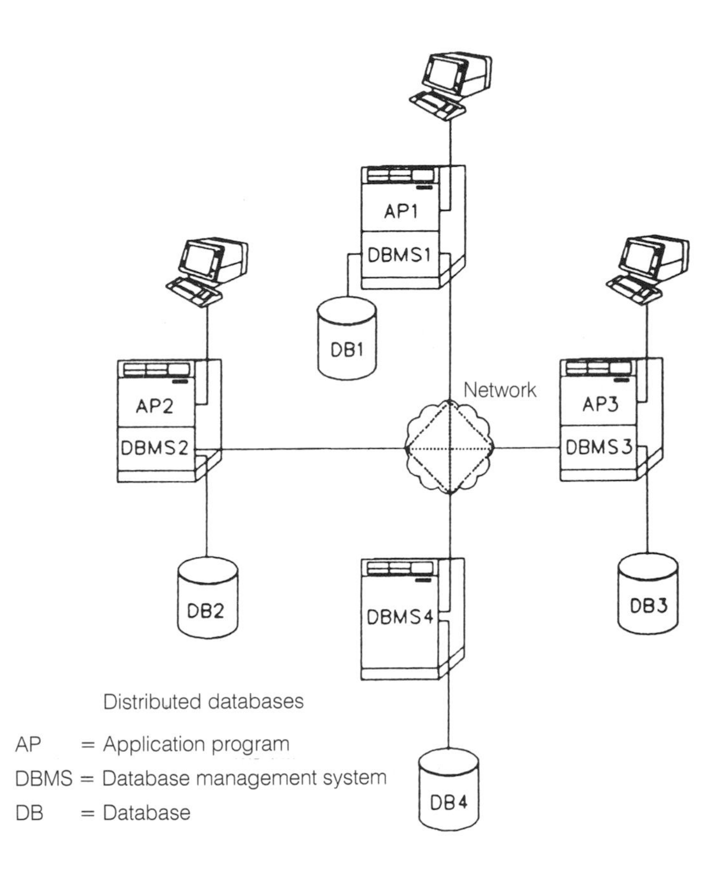

Fig. 2.10 Distributed databases (Effelsberg 1987).

By so-called snapshots, partial copies of the public database can be kept, which have to be kept up to date by periodical updates performed by the local database management system. The workstations with their individual complete database systems and the federal server are connected by a local area network. The main advantage is that at least temporarily the individual workstations can operate autonomously.

2.5 Summary

There are several different types of data exchange between applications.

Relatively simple forms of a data processing connection are the coupling by formatted file or by converters. Both types are relatively widely used, normally as solutions implemented by the users themselves.

With regard to converter processors, extensive efforts aimed at standardization are being undertaken (see Chapter 3).

The high number of necessary intermediate files or converters as well as data redundancy and inconsistency is a major drawback.

A more comprehensive step with regard to integration is connection by databases.

In the area of administrative applications, standardized database systems are applied. The hierarchical model appears not to be practicable. The most advantageous concept is the relational model. For use in practice, means have to be found to accelerate the data access times considerably. In the geometrical/technological sector, standardized database systems do not meet the necessary requirements. A strong research effort has to be undertaken aimed at the development of non-standard database systems.

A central database for all applications involved in computer integrated manufacturing does not exist and does not appear to be useful for a number of reasons.

In the future, distributed database systems, i.e. a combination of a central database and integrated decentralized databases, will be applied (federal database server). The central database stores the public data for all applications. Data private to the individual applications are stored locally. Also in this area much work remains still to be done.

In practise, standardized database systems today are available mainly in the administrative area. In CA applications the file concept is predominant. For the time being, connections between the various applications can be achieved mainly by formatted files or by coupling processors; standardization seem possible.

Chapter 3

Interface standardization

The standardization of interfaces should be based on conceptual models of integrated information and communication systems; as a rule they provide, in the form of a process oriented and abstract scheme, an invariant representation of a system characterized by functions which are connected by information flows. These conceptual models have to be put in a concrete form by giving them a generally valid framework which is independent of the characteristics of individual cases. The results are reference models which represent typical 'standardized' functions and information flows; these can form the basis for the definition of interfaces (Scholz-Reiter 1991). Some recent comprehensive approaches to such models will be discussed in the following chapter without going into detail (for a detailed description and for further models see Scholz-Reiter 1990):

- NBS model of a hierarchical control structure of an integrated, automated manufacturing system (Albus *et al.* 1981, McLean *et al.* 1983);
- CAM-I Model (Computer Aided Manufacturing - International) (CAM-I 1983);
- ICAM architecture (Integrated Computer Aided Manufacturing) (ICAM 1978, ICAM 1981, Harrington 1984);
- Design Rules for a CIM system (Yeomans *et al.* 1985);
- CIM-OSA (Computer Integrated Manufacturing - Open System Architecture) (CIMOSA 1989) (see also Chapter 7).

The standardization of CIM interfaces is still beginning. Merely for the data exchange between CAD systems and the data transfer from CAD systems to functionally related applications (for instance NC part programming), a standardization has been achieved or proposals have been put forward which have a certain degree of relevance to actual practice. For the management of order data there are standardized interfaces only for data exchange between companies. The connection of order oriented and manufacturing related applications necessitates standardizations which have yet to be achieved. In the DIN reports 15, 20 and 21, the necessary effort is discussed and to some extent concretized, especially the aspect of an international standardization in the areas of CAD, NC process chain, production control and manufacturing order processing (DIN 1987c, DIN 1989, DIN 1989a).

In current practice, the standardization of data transfer, i.e. the coupling of applications is more relevant than integration by a common conceptual scheme, i.e via databases.

Interfaces can be placed into two categories: on the one hand there are those for the connection of applications running with different hardware, which for example achieve a portability for the generation and reproduction of graphic elements on different kinds of devices, and on the other hand above all program or data interfaces for the transfer of geometrical and technological information between application systems.

Especially in the area of graphic systems, the standardization of interfaces between different hardware systems is relatively advanced (for example GKS).

For CAD/CAM applications there exists a wide range of alternative standardization proposals for data interfaces, focussing on different functions (IGES, SET, VDAFS, etc.), which address the exchange of product definition data among CAD systems and also data transfer to functionally related applications (NC programming). For the first development step aiming at the coupling of CAD and NC programming systems for the transfer of exclusively geometrical information, the language interface APT (Automatically Programmed Tools) has often been chosen. APT is a universally applicable NC language which is internationally standardized (ISO/TC 184/SC3). Neutral formats such as CLDATA and IRDATA for the exchange of production technology data have to be mentioned as data interfaces for NC and robot programming (see also Section 3.3).

Standardization approaches with regard to the definition of protocols for local area networks which enable company-wide communication of CIM applications are discussed in Chapter 4.

3.1 Interfaces for the connection of applications with graphic devices

The GKS (Graphical Kernel System) group contains standards for interfaces between applications and graphic devices; there are three varieties for two-dimensional graphics, three-dimensional graphics, and for dynamic graphic structures (PHIGS: Programmer's Hierarchical Interactive Graphics System).

GKS-2D has been a German standard since 1986 (DIN 66 252a) and an international standard since 1985 (ISO 7942). In these standards the functions of the graphic systems are defined independent of a programming language. Different application programs can be built around GKS; it can be used not only for CAD programs, but also for simulation, animation, and other graphic application areas. GKS functions can be included in application programs by so-called language shells which are also standardized. On the hardware side - the I/O devices - there is a workstation interface, also called the Computer Graphics Interface (CGI) (ISO 9636). For devices which do not support this interface, a specially designed, complex driver as an additional hardware or software component is necessary which converts the GKS information to the requirements of the individual device.

For long-term storage and the exchange of graphics between the different systems, GKS provides facilities for data input and output to graphics files (GKSM (GKS-Metafile), CGM (Computer Graphics Metafile)); however, only CGS is currently being standardized (ISO 8632, DIN 66 293 draft).

PHIGS (ISO 9592/1-198n(E)) is expected to be standardized by ISO in the near future. PHIGS is not compatible with GKS-3D (ISO 8805, DIN 66 252b, ANSI 1986). A solution could be offered by PHI-GKS, the concept of which ensures compatibility with the GKS-3D environment while offering the functionality of PHIGS (see Noll *et al.* 1987).

For workstations, X-windows provide a standard graphics user interface (ANSI X3H3).

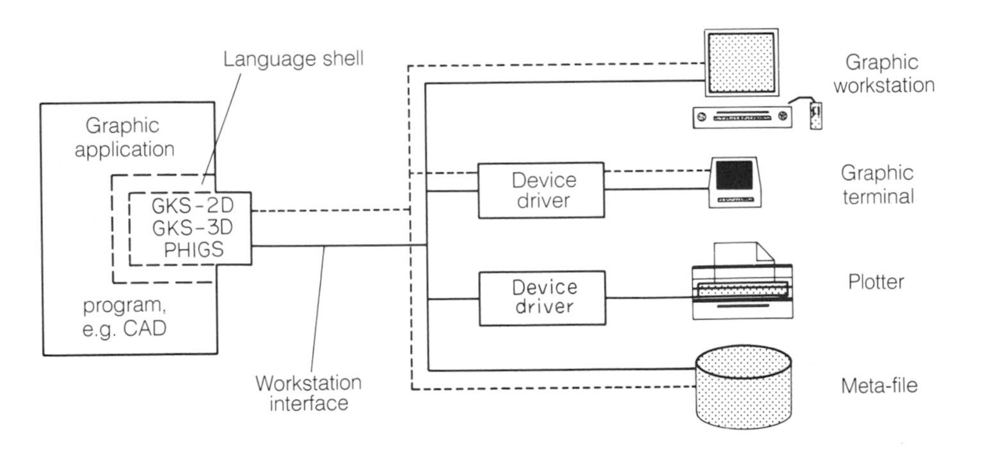

Fig. 3.1 Examples of GKS interfaces.

The above-mentioned standards ensure the portability of graphic applications , i.e. their independence from specific graphic devices. A common exchange format for graphic data or image data does exist.

(For more details on GKS, see for instance Encarnacao *et al.* 1987, Enderle *et al.* 1987.)

3.2 Interfaces for the exchange of product definition data

The issue of interfaces for the exchange of product definition data has been tackled in detail especially by Grabowski and his co-researchers (for the following see, among others, Grabowski *et al.* 1986, Grabowski *et al.* 1987, Anderl 1986a, Anderl 1986b, Anderl 1987, Anderl 1989a, and Grabowski *et al.* 1989).

Product definition data include on the one hand geometrical and on the other hand technological and administrative information. The latter are for example information about materials, tolerances, surface quality, but also information about normed parts and bills of material contained in the design.

Product definition data are generated by the design engineer during the design process and are stored in the CAD system as a computer internal model of the real design object.

Many of the CAD systems currently in use do not permit the description of technological and administrative information. Only the geometrical model of the object is stored. A small number of more modern systems also provide the capabilities of registering and storing technological and administrative information.

The computer internal representations of one and the same object differ widely with regard to the geometrical information, depending on the applications and tasks the CAD system focusses on. For clarification, Fig. 3.2 gives a rough classification of the geometrical elements used for the generation of computer internal models used by CAD systems (Grabowski *et al.* 1986).

For the exchange of product definition data between systems with different computer internal models, coupling processors as described in Section 2.3 are therefore necessary.

For reducing the number of coupling processors and in order to increase flexibility, a wide range of standardization efforts are currently being undertaken, none of which have yet been concluded. The interfaces currently being standardized are mostly geometrically oriented. This is a consequence of the fact that most of the currently applied CAD systems are geometrically oriented, too.

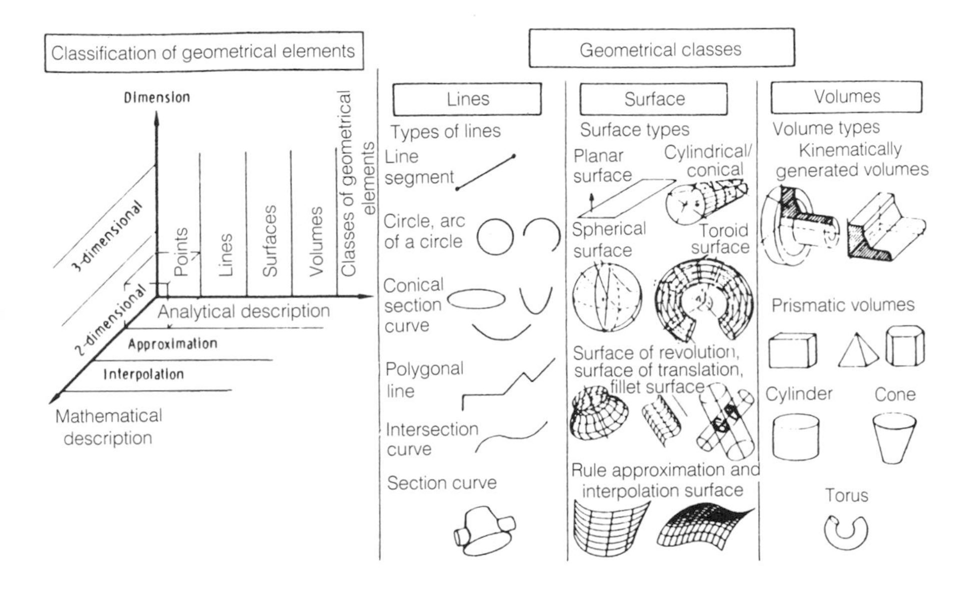

Fig. 3.2. Geometrical elements of varying complexity (Grabowski *et al.* 1986).

This situation is unsatisfactory because manufacturing instructions are generated by the coupling of geometrical, technological, and administrative information. With the transfer of technological and administrative information being excluded, the product model has to be processed once again before the actual manufacturing process can take place. A transfer of this data is useful if corresponding interfaces do exist and new CAD systems are available which also enable the collection and computer internal representation of technological and administrative data.

In contrast to interfaces such as IGES (Initial Graphics Exchange Specification) and VDAFS (VDA Surface Interface) which are mainly geometrically oriented, the proposed interfaces PDDI (Product Data Definition Interface) and PDES (Product Data Exchange using STEP) as well as CAD*I (CAD*Interfaces) are aimed at including the exchange of technological and administrative data.

These interfaces are mainly adapted to the exchange of product definition data between the different CAD systems and between CAD systems and functionally adjacent application areas.

Most standardized interfaces feature a neutral exchange model. The transfer process of the product model is conducted as follows:

The computer internal model of the system A is transformed into the neutral exchange model in the form of an exchange file by a special preprocessor. A postprocessor adapted to the requirements of system B then reads this neutral model and transforms it into the computer internal model of system B. The transfer is completed. This process is performed in the same manner for the transfer between B and A. It is outlined in Fig. 3.3 by Grabowski *et al.* (1986).

The transfer of the exchange file can be done off-line as well as on-line. Because of the large volume of data involved, off-line transfer is generally preferred in practice. The standardized exchange models differ with regard to the number of predefined model elements and the predetermined logical relationships between them. Thus each interface standard focusses on certain tasks. They are not applicable in the same way for the transfer of all computer internal models or even only parts of them, and therefore cannot be universally applied. Moreover, tests have shown that data are lost or altered even when they are being transferred back to the same system. This shows that the interface standards are not yet mature.

In spite of these difficulties, three interface standards are currently being applied in practice:

- IGES,
- SET (Standard d'Echange et de Transfert), and
- VDAFS.

They are the result of national developments (USA, France, and Germany respectively), have different focusses and, with others, are on an international level being incorporated in the specification of STEP (Standard for the Exchange of Product Model Data).

The standardization approaches in the field of the exchange of product definition data, arranged according to the historical development and their place of origin, are shown in Table 3.1. Important standardization efforts are described subsequently.

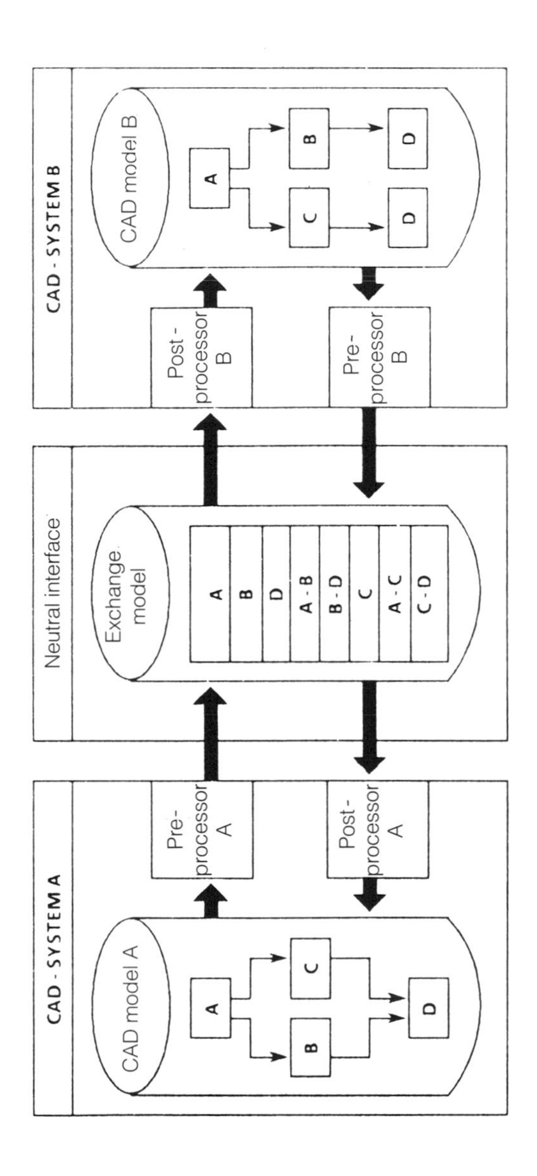

Fig. 3.3 Model exchange between different CAD systems based on neutral interfaces (Grabowski *et al.* 1986).

Table 3.1 Overview of important standardization efforts for the exchange of product definition data

Year	Federal Republic of Germany		France		European Community		USA		International	
	Standard project	Organization	Standard project	Organization	Standard project	Organization	Standard project	Org.	Standard project	Org.
1980							IGES 1.0	NIST		
1981							Y 14.26 M (IGES 1.0)	ANSI		
1982							XBF - 2	CAM - I		
1983	VDAFS 1.0	VDA					IGES 2.0	NIST		
1984			SET 1.1	AEROS			PDDI IGES 2.1	US AIR NIST		
1985	66301 (VDAFS)	DIN	Z 68-300 (SET 1.1)	AFNOR			ESP	NIST		
1986	TAP VDAFS 2.0	DIN VDA			CAD*I 2.1	CEC (ESPRIT-Projekt 322)	IGES 3.0	NIST		
1987	V 66304 (VDAPS VDAIS 1.0	DIN VDA VDA			CAD*I 3.2	CEC (ESPRIT-Projekt 322)	Y 14.26 M (IGES 3.0) EDIF V 200 RS 548 (EDIF V 200)	ANSI EIA ANSI		
1988	V 4001 (CAD-NT)	DIN	SET enlargements (Solid, Scientific Data Finite Elements)	AEROS	CAD*I 3.3	CEC (ESPRIT-Projekt 322)	IGES 4.0	NIST		
1989							PDES 1.0	NIST		
1990										
. . .							IGES 5.0	NIST	STEP(ISO/ TC184/ SC4)	ISO

Legend:

AEROS	AEROSPATIALE
AFNOR	ASSOCIATION FRANCAISE DE NORMALISATION
ANSI	AMERICAN NATIONAL STANDARD INSTITUTE
CAM-I	COMPUTER AIDED MANUFACTURING-INTERNATIONAL
CEC	COMMISSION OF THE EUROPEAN COMMUNITIES
DIN	DEUTSCHES INSTITUT FUER NORMUNG (GERMAN INSTITUTE FOR STANDARDIZATION)
EIA	ELECTRONIC INDUSTRIES ASSOCIATION
ISO	INTERNATIONAL ORGANIZATION FOR STANDARDIZATION
NIST	NATIONAL INSTITUTE OF STANDARDS AND TECHNOLOGY (formerly NBS - NATIONAL BUREAU OF STANDARDS)
U.S.AIR	U.S. AIR FORCE
VDA	VERBAND DER AUTOMOBILINDUSTRIE (ASSOCIATION OF AUTOMOBILE MANUFACTURERS)

Standards:

CAD*I	CAD*INTERFACES
CAD-NT	CAD-NORMTEIL-DATEI (CAD NORMED PARTS FILE)
EDIF	ELECTRONIC DESIGN INTERFACE FORMAT
ESP	EXPERIMENTAL SOLID PROPOSAL
IGES	INITIAL GRAPHICS EXCHANGE SPECIFICATION
PDDI	PRODUCT DEFINITION DATA INTERFACE
PDES	PRODUCT DATA EXCHANGE SPECIFICATION (originally) PRODUCT DATA EXCHANGE USING STEP (now)
SET	STANDARD D'ECHANGE ET DE TRANSFERT
STEP	STANDARD FOR THE EXCHANGE OF PRODUCT MODEL DATA
TAP	TRANSFER UND ARCHIVIERUNG PRODUKTDEFINIERENDER DATEN (TRANSFER AND STORAGE OF PRODUCT DEFINITION DATA)
VDAFS	VDA-FLAECHENSCHNITTSTELLE (VDA SURFACE INTERFACE)
VDAIS	VDA - IGES SUBSET
VDAPS	VDA-PROGRAMMSCHNITTSTELLE (VDA PROGRAM INTERFACE)
XBF	EXPERIMENTAL BOUNDARY FILE

3.2.1 IGES (VDAIS, PDES)

In 1980, IGES-Version 1.0 (Initial Graphics Exchange Specification) as the first standard for the exchange of model data was published by the NBS (National Bureau of Standards, now: NIST - National Institute of Standards and Technology). In 1981, this version was included in the ANSI (American National Standard Institute) standard Y14.26M (ANSI 1981).

Since then the IGES committee, consisting of a team of more than a hundred people, has continuously been working on the further development of IGES:

- IGES-Version 2.0 (1983);
- IGES-Version 2.1 (1984);
- IGES-Version 3.0 (1986);
- IGES-Version 4.0 (1988); and
- IGES-Version 5.0 (after 1990).

For this reason, of all interfaces for the exchange of product definition data, IGES is currently the most advanced.

IGES 1.0 was adapted to the CAD systems as they were used throughout the 1970s; its concept aimed at the data exchange between CAD systems. This also explains why later versions of IGES mainly supported the transfer of geometrical information of computer internal models.

Currently Version 2.0 (IGES 1983), next to Version 3.0, is the basis for most of the applied and marketed IGES processors. This version is subsequently presented.

In order to assure independence from specific hardware and the readability of the neutral IGES file, a simple physical file format was chosen. It is sequentially organized with a record length of 80 characters using the ASCII character set. Because of this the amount of required storage space is necessarily high (about three to five times as much as the CAD internal format requires), for which reason the transfer in a binary format (and, since IGES 3.0, also in a compressed ASCII format) was made possible. This, however, is rarely done by the average user as the data are difficult to read and interpret. The binary format plays an important role mainly for the transfer of IGES files within computer networks.

Logically an IGES file is divided into five predetermined sections (from IGES 3.0 onwards a FLAG section may precede them which indicates whether the file is of the compressed ASCII or binary type, if it is not a standard file):

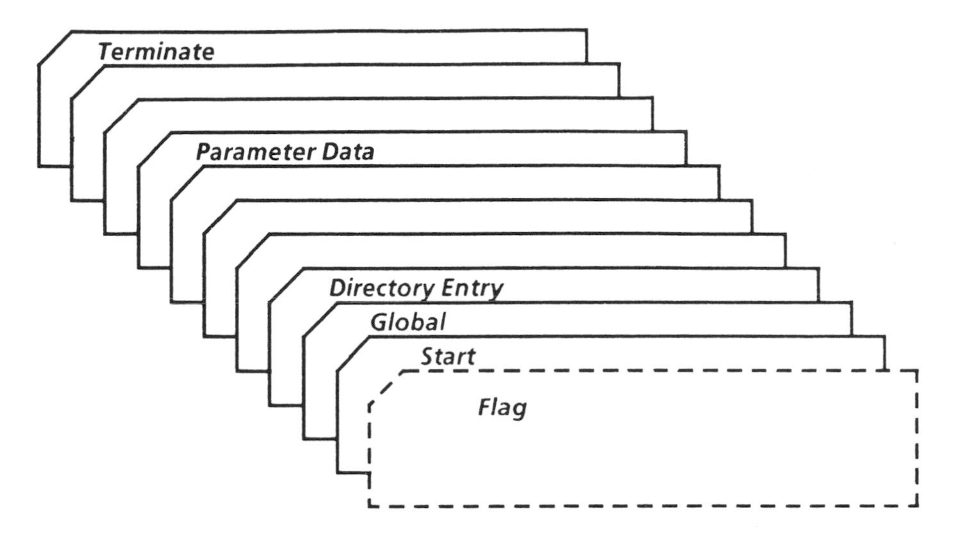

Fig. 3.4 Logical structure of an IGES file.

- The START section contains comments and messages to the receiving system.
- The GLOBAL section contains information for pre-setting the system of the postprocessor, depending on the preprocessor used.
- In the DIRECTORY ENTRY section, all elements (e.g. geometrical elements) specified within the IGES file are laid down, independent of the actual characteristics of the element (two data sets in a fixed format for each element).
- In the PARAMETER DATA section, the individual characteristics of all the previously specified elements are stored in a free format.
- The TERMINATE section serves as an end-of-file marker and contains the number of strings making up the several file sections as a test sum.

Within the IGES file, the product definition data are represented as a list of information elements, called entities, which may be logically connected to each other by pointers. As the previous description shows, the entities are all contained in the DIRECTORY ENTRY and the PARAMETER DATA section which are connected to each other by pointers.

The element types can either be predefined elements or elements specifically defined by the user. The predefined elements are used for describing the geometrical shape of an object (geometry entities), the dimensions and technological information about an object within a design (annotation entities), and for describing the logical structure of the product model (structure entities).

In addition to that, the IGES user can define individual elements which are not considered by IGES (user defined elements). As these are individual definitions, the whole structure has to be known to both the preprocessor and the postprocessor in order to guarantee an orderly interpretation during the transfer. Therefore user defined entities are rarely used.

With the help of macros (similar to BASIC), shape variable geometries, i.e. parameterizable elements, can be transferred easily.

In Fig. 3.5, an IGES file for the drawing of a simple outline of a metal sheet is listed. The IGES file for the subsequent drawing of a piston consists of about 1900 lines.

Apart from 2-D drawings (wireframe models), 3-D models (wireframe and surface models) can also be transferred by an IGES interface.

While IGES 1.0 was mainly oriented towards the transfer and storage of product data from the mechanical engineering sector, IGES Version 2.0 has been enlarged by the electrical engineering and electronics application areas, and allows the transfer of finite element models and rational B-spline surfaces.

In Version 3.0 (IGES 1986), the following improvements were made:

- a more precise specification, i.e. implementation guidelines for processors;
- compressed ASCII format for reduction of the required memory space (to about one third);
- revised element definitions;
- enlarged number of elements, especially for annotation elements; and
- enlargement of macros and additional control structures (cf. Grabowski *et al.* 1986).

Version 3.0 replaced IGES 1.0 as the ANSI standard.

Version 4.0 additionally enabled the transfer of solid models on the basis of ESP (Experimental Solid Proposal) (ESP 1984). ESP itself is based on XBF (Experimental Boundary File) (XBF 1981) which was developed by CAM-I (Computer Aided Manufacturing - International). XBF represents solid models by boundary surfaces (CSG - Constructive Solid Geometry). It contains 44 geometrical and 11 topological IGES-compatible elements. For the support of application areas such as the finite element method (FEM) and electronics, special non-geometrical elements were introduced.

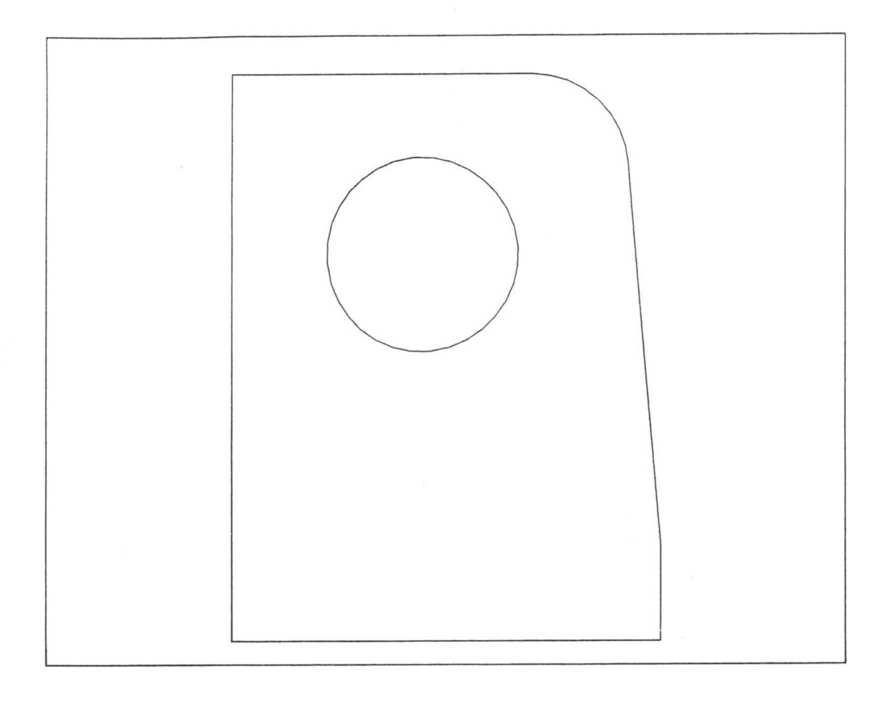

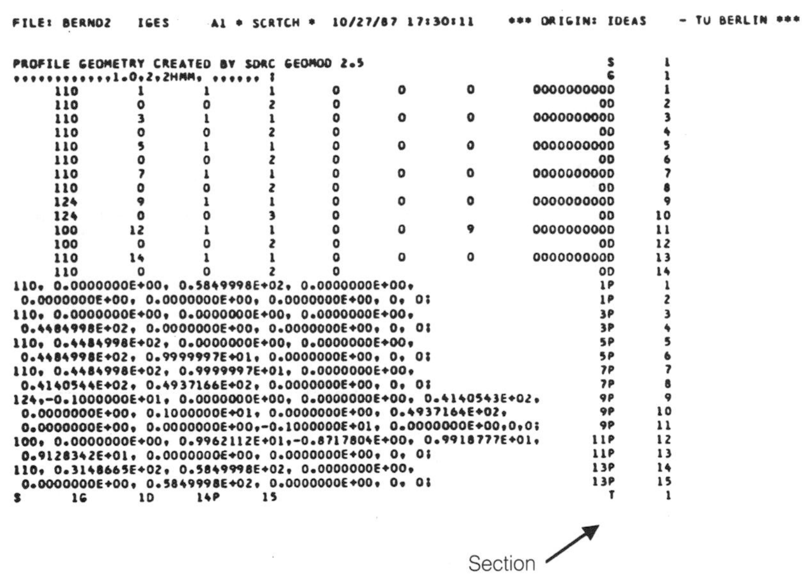

```
FILE: BERND2   IGES    A1 * SCRTCH *  10/27/87 17:30:11    *** ORIGIN: IDEAS    - TU BERLIN ***

PROFILE GEOMETRY CREATED BY SDRC GEOMOD 2.5                                   S      1
,,,,,,,,,,,,1.0,2,2HMM, ,,,,,, ;                                              G      1
     110       1       1       1       0       0       0        000000000D      1
     110       0       0       2       0                               0D      2
     110       3       1       1       0       0       0        000000000D      3
     110       0       0       2       0                               0D      4
     110       5       1       1       0       0       0        000000000D      5
     110       0       0       2       0                               0D      6
     110       7       1       1       0       0       0        000000000D      7
     110       0       0       2       0                               0D      8
     124       9       1       1       0       0       0        000000000D      9
     124       0       0       3       0                               0D     10
     100      12       1       1       0       0       9        000000000D     11
     100       0       0       2       0                               0D     12
     110      14       1       1       0       0       0        000000000D     13
     110       0       0       2       0                               0D     14
110, 0.0000000E+00, 0.5849998E+02, 0.0000000E+00,                             1P      1
 0.0000000E+00, 0.0000000E+00, 0.0000000E+00, 0, 0;                           1P      2
110, 0.0000000E+00, 0.0000000E+00, 0.0000000E+00,                             3P      3
 0.4484998E+02, 0.0000000E+00, 0.0000000E+00, 0, 0;                           3P      4
110, 0.4484998E+02, 0.0000000E+00, 0.0000000E+00,                             5P      5
 0.4484998E+02, 0.9999997E+01, 0.0000000E+00, 0, 0;                           5P      6
110, 0.4484998E+02, 0.9999997E+01, 0.0000000E+00,                             7P      7
 0.4140544E+02, 0.4937166E+02, 0.0000000E+00, 0, 0;                           7P      8
124,-0.1000000E+01, 0.0000000E+00, 0.0000000E+00, 0.4140543E+02,              9P      9
 0.0000000E+00, 0.1000000E+01, 0.0000000E+00, 0.4937164E+02,                  9P     10
 0.0000000E+00, 0.0000000E+00,-0.1000000E+01, 0.0000000E+00,0,0;              9P     11
100, 0.0000000E+00, 0.9962112E+01,-0.8717804E+00, 0.9918777E+01,             11P     12
 0.9128342E+01, 0.0000000E+00, 0.0000000E+00, 0, 0;                          11P     13
110, 0.3148665E+02, 0.5849998E+02, 0.0000000E+00,                            13P     14
 0.0000000E+00, 0.5849998E+02, 0.0000000E+00, 0, 0;                          13P     15
S      16      1D      14P      15                                            T      1
```


Fig. 3.5 IGES file of the simple outline of a metal sheet.

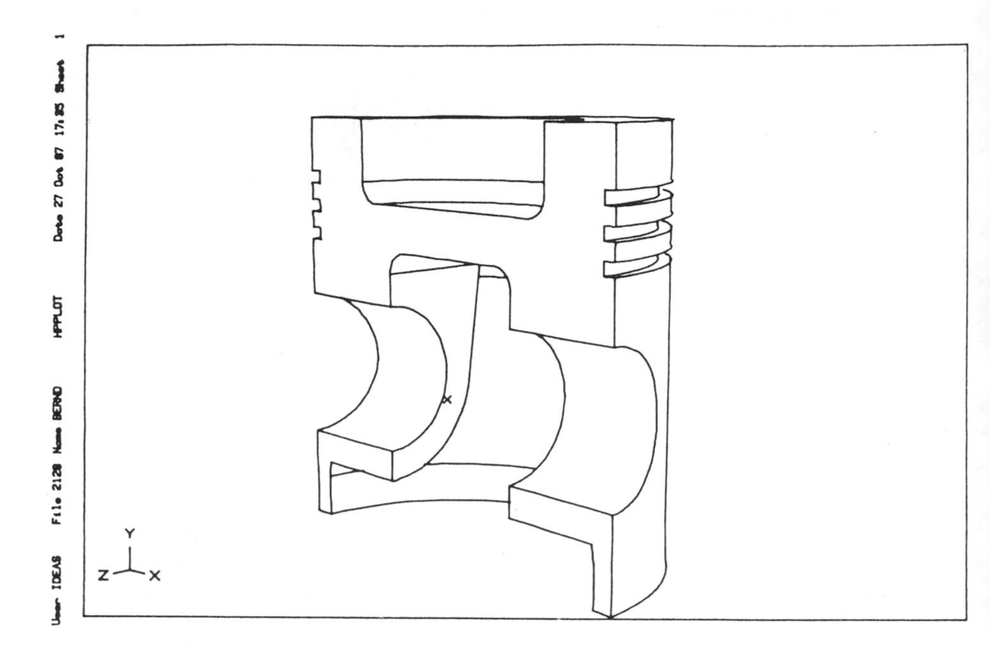

Fig. 3.6 CAD drawing of a piston.

For IGES 5.0, it is planned to allow the transfer of solid models with boundary representation (B-Rep). After that the development of IGES will cease or be integrated into the STEP development.

In order to enable the transfer of complete product models in the future, the IGES committee is conducting the specification of a wholly new interface called PDES (originally: Product Data Exchange Specification, now: Product Data Exchange using STEP) which is no longer mainly geometrically oriented, but also allows the complete transfer of technological and administrative information (see PDES 1990a and Section 3.2.6).

Functionally, the first valid PDES version will completely include the valid IGES Versions 4.0 or 5.0. However, the physical and logical structure of these two neutral interfaces will be completely different. For the transfer of IGES files into the PDES format, converter programs will be implemented.

Above all, in addition to the IGES functions, functions for the transfer of technological and adminstrative information will be implemented, such as they have been formulated in the framework of the PDDI project (Product Definition Data Interface, see section 3.2.4). PDES forms an essential basis for the international standard STEP (Standard for the Exchange of Product Model Data, see Section 3.2.6).

Currently, IGES processors for the Versions 2.0 and 3.0 are mainly being applied. Even when using them within the limits of their specified functions, the following problems frequently do occur:

- incorrect transfer of geometry and drawing elements such as dimensions;
- structural elements are interpreted differently by the processors;
- only a subset of the data elements available within the CAD system is transferable (Schwindt 1986);
- no test of the syntax is conducted in advance of the conversion process, which leads to termination or wrong results;
- terminations occur without specific error messages;
- there is no indication of which data have been transferred correctly, only approximatively, or not at all (Bey *et al.* 1989).

IGES is well suited for the tansfer of structured drawing information. The transfer of more elaborate product models which must be available in the form of structured geometrical models is often problematic with regard to a complete data exchange without losses, resulting from the existing latitude concerning the implementation of processors (cf. Nowacki 1990, Grabowski *et al.* 1989). The Verband der Automobilindustrie (VDA - Association of Automobile Manufacturers) has therefore determined a subset of IGES Version 3.0 as a minimal scope for the interface VDAIS (VDA-IGES subset) (VDMA/VDA 1987) which includes mainly the drawing oriented elements and the free-form geometry relevant for automobile construction. Syntax and translation errors are to be intercepted or kept in a journal.

There exist international and national working groups and institutions for the test and verification of IGES processors.

3.2.2 SET

As a counter-proposal to IGES with the aim of avoiding IGES's deficiencies, SET (Standard d'Echange et de Transfert) was developed in France by Aerospatiale. The first version, 1.1, was published in 1984 and registered as the national standard Z 68 - 300 (AFNOR 1985) by the French institution for standards AFNOR (Association Française de Normalisation). The physical file

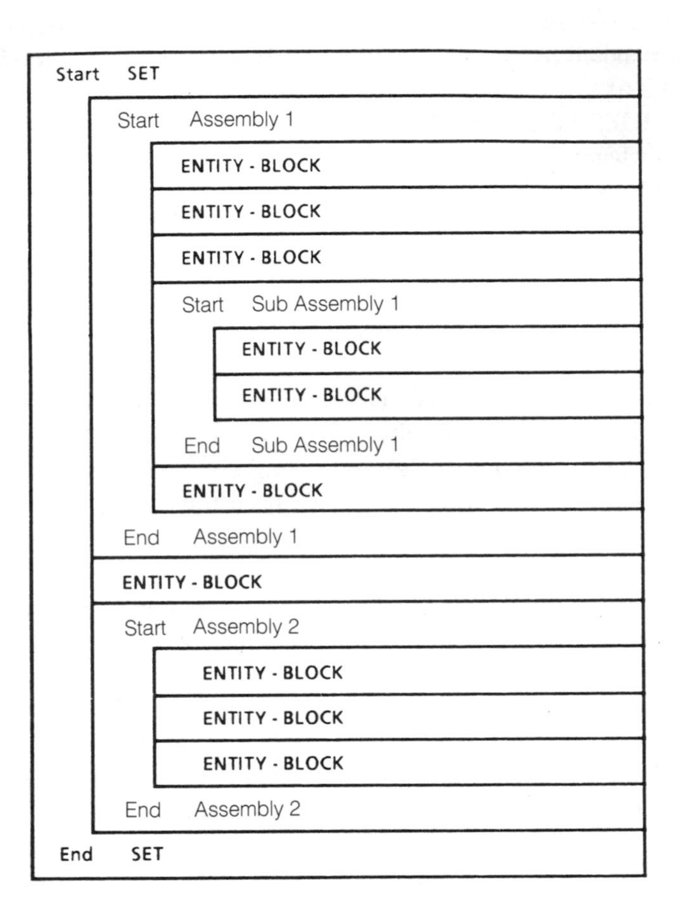

Fig. 3.7 Example of the logical structure of a SET file.

format of a SET file differs from that of an IGES file in that the length of single data records is variable. The end of a record is determined by certain end characters. The logical structure is shown in the Fig. 3.7.

The file consists of blocks which are nested within each other. The compulsory outer block is delimited by a START SET and an END SET card. In this block, one or several entity blocks can be nested which can contain the defined parameters, and/or one or more assembly blocks which themselves may again consist of one or more entities and/or assembly blocks, delimited by START and

END cards and logically representing single parts or assemblies. There is a multiple level depth of nesting.

In addition, there is a dictionary in which certain parameters can be initialized globally. If initialized parameters are revised within one block, this change is valid only for this one block. In contrast to IGES files, the SET file structure has no reverse pointers. Macro definition is not available.

As a consequence of the described concept, the required memory space is smaller and processor run-times are more advantageous than in the case of the IGES format.

SET is also intended to allow the complete storage of all computer internal model data of different CAD systems in a generally valid external database. SET in the first place supports mechanical engineering. The available elements therefore are more versatile than those provided by the applied IGES Versions 2.0 and 3.0. In 1988, enlargements were made for solid models, FEM, and scientific product data (Nowacki 1990).

SET supports the tranfer of 2-D and 3-D wireframe models, 3-D surface models and solid models (CSG, B-Reps, facette models).

SET is above all being applied by the European aerospace industry for the Airbus development. Via AFNOR, SET contributes to the international standardization efforts undertaken by ISO.

3.2.3 VDAFS

Especially for the exchange of free-form curves and surfaces without restrictions in degree, to which IGES is only partially suited (up to the third degree), VDAFS (VDA Surface Interface) was conceived by VDMA (Verband Deutscher Maschinen- und Anlagenbau, Association of German Machine and Plant Manufacturers) (VDA/VDMA 1983). It was introduced in 1983, and in 1986 it was nationally standardized as DIN 66 301. A further Version 2.0 exists in the form of a VDA recommendation (Mund *et al.* 1987).

In the same way as an IGES file, the VDAFS file is structured sequentially with a fixed record length of 80 characters using the ASCII character set.

Currently, due to its narrowly limited scope, this interface contains no more than 5 geometrical entities (single point, point sequence, point sequence with direction sectors, curve in polynomial representation and surface in polynomial representation). In Version 2.0, these have been enlarged to circle arc, outline on surface, surface with boundaries, and surface formation.

Further data elements serve for marking beginning and end (header, end), the formation of the element groups (beginset, endset), and as comment markers ($$).

```
VDAFS     = HEADER /11                                                          00000005
*********************************************************************************00000010
 1. Senderfirma              BROSE FAHRZEUGTEILE                                 00000015
 2. Projekt                  TEST                                                00000020
 3. Dateiname                TEMP VDA2
 4. Gueltigkeitsdatum        24 12.85                                            00000030
 5. Erzeugungsdatum          25 2 86 18 34 58
 6. Erzeugendes System       CADDS 4x
 7. Ansprechpartner          H. GUELERYUEZ
 8. Telefonnummer            09561/21-453
 9. Adresse                  8630 COBURG
*********************************************************************************00000060
SET00001 = BEGINSET                                                              00000065
CP000001 = CURVE   /1.0.1..2.117.739,0..0.3.-87.77377,0.                        00000070
CP000002 = CURVE   /1.0.1..2.26.7858,90.96576,0..0.-8.709396,                   00000075
                    -79 07533                                                    00000080
CP000003 = CURVE   /1.0.1..2 0.8299054,1.11924,0.0..-149.526,                   00000085
                    -3.233353                                                    00000090
CP000004 = CURVE   /1.0.1 2.15.21448,-0.2372589,0.0..-19.8508,                  00000095
                    0 2062454                                                    00000100
CP000005 = CURVE   /1.0.1 2.14 9737.0 000001907349,0..0.-19 6415,               00000105
    .               -0 000001907349                                              00000110
    .
    .
    .
CP000364 = CURVE   /1.0.1..8.0.1997605,0.2936382,0 6970056,-0 5038344,          00003815
                    -0.08752652,0.1064606,-0.0189773,0.00071 0994.               00003820
                    -0.000001697954,-4 94243,3.770065,3.532892,-2.75666,         00003825
                    0.3358511.0.0721 8228 -0.01189682,22.14069,2.391497,         00003830
                    5 676619,-4 103298,-0 7129669,0 8671761.-0 1546898,          00003835
                    0 005849838                                                  00003840
CP000365 = CURVE   /1.0.1..4 5.406231.2.533459,-0.2218108,-0.3265567,           00003845
                    -1.25,0 0000007152557,1.45563,-0.2056305,23.51644,           00003850
                    -0.3110676,0.0272322,0.0400981 9                             00003855
CP000366 = CURVE   /1.0.1..4.-1.541595,1.11017,1 201478,-0.3265553,             00003860
                    -0.00000000825476 5,-2.294369,0 8387389,0 20563,             00003865
                    24.36952,-0.136306,-0 1475258,0.04009819                     00003870
CP000367 = CURVE   /1.0.1..8 32 5.0.-62.94985.57.6991 8.141 651 7.              00003875
                    -204.6016.68 20052.0..0..0.0..0..0..0..0..0.-25..            00003880
                    17 74241.-8 179333.-132 7239.165 325.17 75023,               00003885
                    -83 88024 23 96578                                           00003890
CP000368 = CURVE   /1.0.1..8.32 5.0.-62.94985.57.6991 8.141 651 7.              00003895
                    -204.6016,68 20052.0.3..0..0.0..0..0..0..0.-25..             00003900
                    17.74241.-8.179333.-132.7239.165.325.17.75023,               00003905
                    -83 88024.23 96578                                           00003910
SET00001 = ENDSET                                                                00003915
SET00002 = BEGINSET                                                              00003920
P0000001 = POINT   /-21 36826 0..-4.                                             00003925
P0000002 = POINT   /-14 9737.0.-19 6415                                          00003930
P0000003 = POINT   /14 9737 0.-19 6415                                           00003935
P0000004 = POINT   /-11 .0 30                                                    00003940
P0000005 = POINT   /11..0,30                                                     00003945
SET00002 = ENDSET                                                                00003950
VDAFS     = END                                                                  00004010
```

Fig. 3.8 Example of a VDAFS file (Schwindt 1986).

In Version 2.0, these have been enlarged to header with structure, group formation, transformation matrix and a list of elements to be transported. A characteristic feature is the APT oriented data format.

Up to the present, VDAFS has been consciously restricted to the mere transfer of geometrical data. Therefore, the following restrictions and deficiencies occur in practice:

- text, dimensions, hatching, etc., cannot be transferred;
- mathematical geometry (for instance circles) has to be tranformed into free-form geometry;
- single elements arrive at the receiver in an atomized form and can only be processed with large effort;
- elements for structuring the data exist only in a rudimentary form (Schwindt 1986).

For these reasons, in addition to the data transfer, the original hardcopy drawings are still indispensable. Some of the deficiencies mentioned above have been removed in Version 2.0.

The VDA surface interface has been successfully applied within the limits of its restricted scope by German automobile manufacturers and their parts suppliers. Like IGES processors, VDAFS processors can be examined with regard to their compliance to the standard by a neutral institution. VDAFS has been conceived as a pragmatic intermediate solution which can also contribute to an overall international standardization and will in the future be replaced by it.

3.2.4 PDDI (EDIF)

PDDI (Product Definition Data Interface) was developed within the framework of the ICAM project of the U.S. Air Force and first introduced in 1984 (Weiss 1984).

In contrast to the previously discussed interface standardizations which emphasize data transfer between CAD systems, the main aim of PDDI lies in the transfer of integrated product models from CAD systems to adjacent CA applications, especially to the manufacturing area.

EDIF (Electronic Design Interchange Format - ANSI RS 548) has a similar purpose in the electrical engineering/electronics sector, i.e. the geometrical and technological description of two-dimensional schematic layouts, as for instance of integrated circuits. EDIF, as for example also VNS (Verfahrensneutrale Schnittstelle - method independent interface), an interface for the transfer of circuit layouts, will not be considered in more detail because of their orientation towards the electrical engineering/electronics sector which is not within the scope of this book (for further information see for instance Abel *et al.* 1987, DIN 1989).

Because of its different scope, the PDDI interface concept and contents differ from those mentioned before. The exchange of drawings is not supported; exclusively digital data are transferred, i.e. all manufacturing information contained in construction drawings are digitally represented. In this respect PDDI responds more to the requirements of continuous data flow between computer applications in CIM than the previously discussed concepts.

The exchange files are again sequentially structured and have an ASCII format.

PDDI is oriented towards mechanical products. The sorts of elements essential for the relevant information transfer at the design-manufacturing interface are (Grabowski *et al.* 1986 with reference to Weiss 1984):

- the number of curves, surfaces, and volumes which define the three-dimensional shape of a product (geometry);
- the number of relationship elements that define which elements delimit the product (topology);
- the number of elements which define the acceptable shape deviations (tolerances);
- the number of elements which explicitly describe manufacturing related form elements (features), and
- the amount of additional, non-geometrical information such as materials and part numbers.

PDDI was implemented only as a pilot version with the aim of verifying the underlying concepts, and is not being applied practically. Its importance lies mainly in its influence on the development of PDES and STEP (see Section 3.2.6).

3.2.5 CAD*I

Within the framework of the European Community sponsored ESPRIT project 322 entitled CAD*Interfaces which commenced in 1984, efforts aimed at the development of standardized interfaces have been made (Schlechtendahl 1988, Schlechtendahl 1989a, Bey *et al.* 1986, Bey *et al.* 1988, Bey *et al.* 1989, and Maanen *et al.* 1990). The emphases of this project are:

- data exchange between CAD systems (via networks);
- a neutral database for storing CAD data originating from a wide range of different CAD systems, and
- data exchange between CAD and FEM applications.

The work is conducted in concert with the international STEP development and, together with PDES, form its main input. Preprocessors and postprocessors for 13 different CAD systems are being developed (Schlechtendahl 1989).

Following an analysis of existing standardized interfaces (see preceeding chapters) and new CAD/CAM systems, standardized interfaces have been developed and tested in co-operation with national and international standardization boards with the aim of fulfilling the following requirements (Bey *et al.* 1986):

- representation and exchange of solid, surface, and wireframe models, and finite element models;
- the storage and retrieval of these models using CAD/CAM systems;
- the exchange of such models via networks;
- the application of advanced modelling techniques;
- the standardized coupling of FEM programs;
- the comparison of data gained experimentally and analytically for structural analysis;
- the dynamic optimization of models based on experiments and analyses.

The neutral format CAD*I developed in this project is characterized as follows (Bey *et al.* 1989):

- based on a CAD reference model;
- strictly sequential files without forward pointers, which therefore can be executed in a single operation sequence;
- block structure;
- formal description of the elements by the data structure specification language HDSL (High Level Data Specification Language);
- fixed rules for the translation from HDSL to Bachus Naur form of the description of the physical data format (with reference to compiler design);
- definition of the processor semantics based on the concept of the finite automaton.

In 1987, volume models with boundary representation (B-Reps) could be transferred for the first time with the help of CAD*I (implementation is planned for IGES only from Version 5.0 onwards).

3.2.6 STEP

STEP (Standard for the Exchange of Product Model Data) (STEP 1988) is the unofficial designation of a future international standard which will define an external representation of a product model for all the phases of the entire product life cycle. STEP aims at achieving long-term accessability and intelligibility, completeness and integrity, and exchangeability between different systems. The STEP development is mainly influenced by the previously discussed standards and standardization projects CAD*I and PDES.

A product model in the first place creates unified semantics for all applications participating in product oriented information flows. STEP therefore is not merely a data exchange format, but a complex information model oriented towards the object 'product'. This product model contains mainly topological/geometrical, technological and administrative product data. It is subdivided into corresponding partial models.

A requirements catalogue by ISO (ISO 1985) listing 60 points serves as a framework for the technical development efforts which are mainly conducted by national standardization boards (see also the preceding Sections 3.2.1-3.2.5).

The concepts discussed for the structure of STEP are described by Grabowski *et al.* (1989) as follows:

- a three-layer architecture for information representation (application layer, logical layer, physical layer);
- the use of reference models for a consistent development and definition of the standard;
- the use of formal languages (EXPRESS) (ISO 1990) for the exact and data-processable description of the different layers of the standard; and
- the structuring of the standard in predefined implementation levels, in order to support the implementation and certification of functional subsets of the standard (Grabowski *et al.* 1986).

Fig. 3.9 gives an outline of the structure of the planned STEP standard (for the following paragraph cf. Grabowski *et al.* 1989, Nowacki 1990, Schlechtendahl 1989b, STEP 1990b).

For the relevant application areas (mechanical engineering, construction engineering, electronics, FEM analysis), the application protocols contain corresponding information models (topical information model - TIM) which can also consistently overlap. These application models can be interpreted, i.e. decomposed into application independent terms of the STEP product description. All independent terms of the product description are combined in the integrated product information model (IPIM). The aim is to avoid conflicts and redundancies. The integrated product information model consists of the product definition resource model which contains all commonly applicable information models for material characteristics, form description types, presentation, etc., and the product representation resource model which comprises basic models for

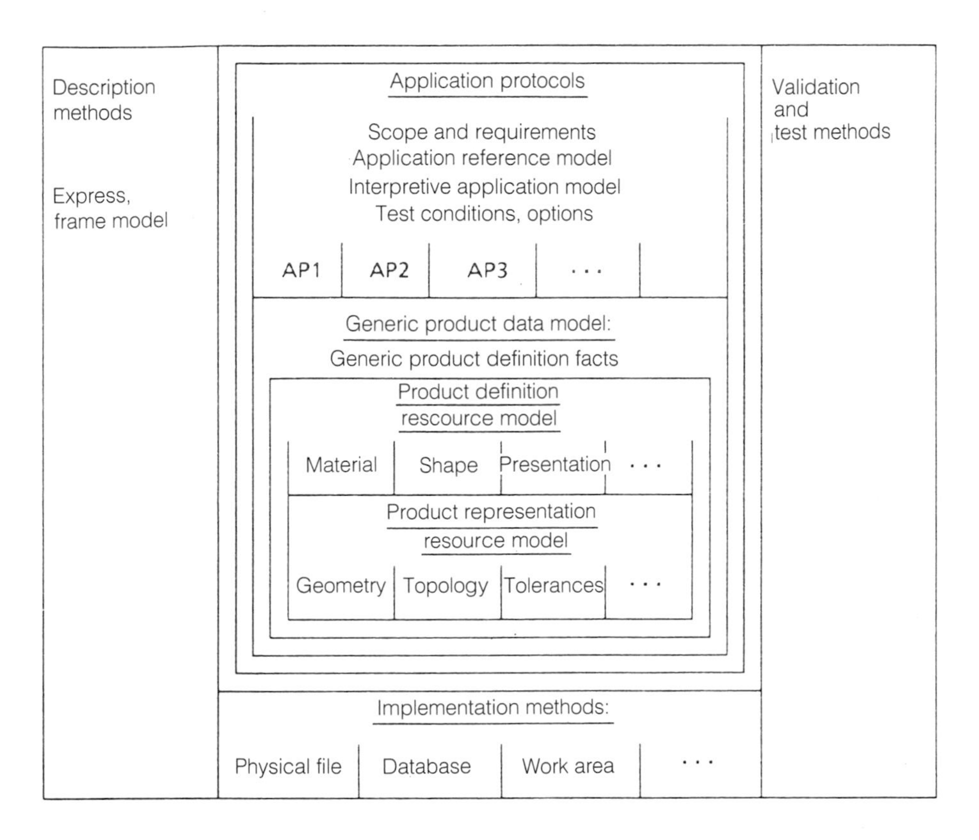

Fig. 3.9 Structure of the planned STEP standard (Nowacki 1990 with reference to Danner 1990).

product representation such as geometry, topology, etc. So far the geometry and topology models can be regarded as mature.

First STEP applications on a prototype level are expected after 1991 (cf. Fowler 1990).

3.2.7 VDAPS and CAD-NT

Since 1984, efforts have been undertaken in the former Federal Republic of Germany to develop standardized interfaces for the provision and the exchange of standardized and repeat parts for and between different CAD systems. VDAPS

(VDA Program Interface) was developed by DIN and VDA and published in June 1987 as the preliminary DIN standard 66 304 (DIN 1987).

By analogy with the conventional documentation of standardized parts by type drawings and dimension tables for the various versions, a procedural approach was chosen with which the special variants of a standardized part are generated with the help of variant programs (in FORTRAN 77). These contain the geometrical generative logic (currently 2-D and 3-D wireframe models) and work with dimension files representing special dimensions in the form of CAD characteristics lists.

VDAPS thus contains the generative logics of standardized parts. The dimension files with which these generative logics operate must themselves be standardized, too. For this purpose DIN defined the CAD-NT file (CAD standard part file) which represents the different variants of standardized parts with CAD appropriate characteristics (DIN V 4001) (DIN 1987b). CAD standard part files as well as the accompanying (variant) geometry programs are marketed by the DIN-Software GmbH.

As a consequence of the concept of the variant programs as passive programs which neither support user communication nor access to the characteristics files for the identification of a single part in the parts family, in advance of running the variant programs the characteristic values of a special standardized part have to be gathered in characteristics files (standard parts tables), which form the input file for the variant program. For this process special methods have been developed. A transfer format was defined for the exchange of characteristics files (cf. DIN 1987c) which is based on the ISO format SQL (Standard Query Language).

The different CAD systems have to be adapted to VDAPS and CAD-NT by the suppliers or software firms. A program for reading and processing the standard parts tables and an adaptor module for the geometry programs (the realization of VDAPS in accordance with DIN V 66 304) must be integrated into the CAD system.

By the end of 1988, 24 standard CAD systems on the German market offered this interface (Kölling 1989). However, the DIN standard parts in the CAD system have to be supplemented by supplier specific data if they are to be used as information for bills of material at a later stage.

Efforts are under way to enlarge the system with 3-D surface and solid models. The relevance of a standardized transfer of standard parts geometries has been realized also by European and international boards; therefore VDAPS and CAD-NT have an influence also on STEP.

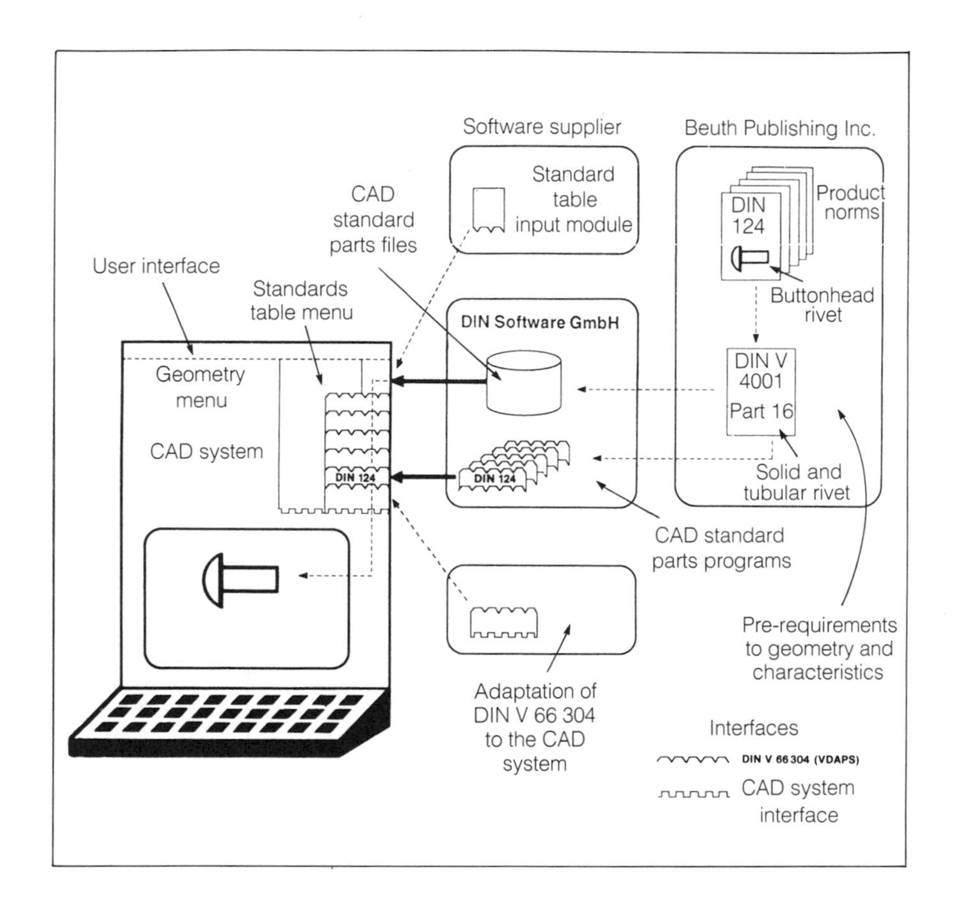

Fig. 3.10 Provision of files and geometry programs for CAD standard parts (courtesy of DIN Software GmbH).

3.3 Interfaces for the exchange of production technology data

Interfaces for the exchange of production technology data are necessary for the automated programming of NC machines and industrial robots (IR). They form a uni-directional link between CAPP and CAM.

In the framework of CAPP, the automated programming of NC machines is done with the help of NC programming systems. The result of this programming are NC data which, standardized as CLDATA (Cutter Location Data) (DIN 66215/ ISO 8632), can be output independently of the machine used. A similar concept applies to robot programming, where for the output of neutral robot control data the VDI guideline 2863, IRDATA (Industrial Robot Data), has been stipulated.

The coding of NC machines itself has been standardized by DIN 66 025. For the coding of industrial robots, no corresponding standardization effort has been undertaken up to the present.

For clarification, these interfaces for the automated NC programming are presented in Fig. 3.11.

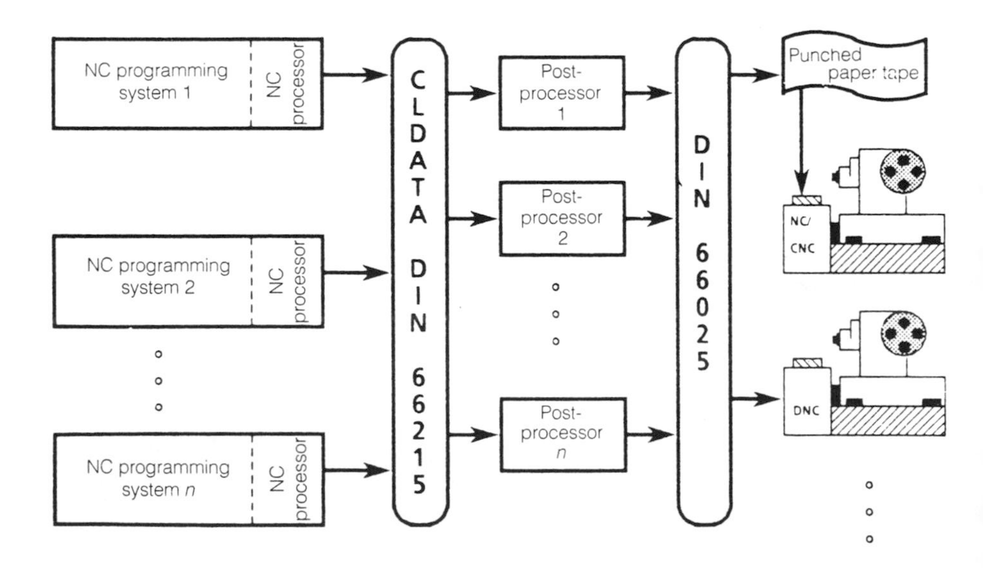

Fig. 3.11 Interfaces for the exchange of production technology data in the case of automated NC-programming.

In the NC programming system, the part program is encoded by the NC programmer in interaction with the system, usually in a formal language (APT, APT dialects, etc.). APT (Automatically Programmed Tools, ISO TC 184/SC3) as a language interface is also used for transferring geometrical data from a CAD system to the NC programming system. In the next step the generated source programs are examined by the NC processor with regard to their formal correctness. If no errors occur, a neutral exchange format (CLDATA) is generated.

CLDATA is a format for NC processor output data which are used as input data for NC postprocessors. Each NC processor - which is a program describing compilation, translation and related functions for a programming language for numerically controlled machines - should be able to generate CLDATA code with a standardized, predefined structure. Each NC postprocessor - which is a program describing the conversion of CLDATA text into specific control data for NC machines - should be able to process CLDATA text in a standardized, predefined form (cf. DIN 1974).

The advantages of this approach are evident. The most diverse NC programming systems can be applied; the results are the same. NC data are generated which are not machine oriented, but part oriented. The same CLDATA text can be used for a number of different NC machines with the help of special postprocessors. This is of special relevance when shifting orders to other machine units (cf. Eigner *et al.* 1985). The flexibility of production control is thus considerably enhanced. When changing over to other NC controllers, the parts programs do not have to be generated once again by an NC programming system.

The program structure for numerically controlled machines has been defined by DIN 66 025 in a standardized form. The NC program is afterwards supplemented by machine specific code.

The data making up the control program are usually stored on data carriers and are input into the controller of the machine in the form and the functional meaning defined by the standard; they can be displayed or output by the controller.

The specifications of the standard are usually not sufficient to allow the direct exchange of the control programs between NC machines of the same class. However, such an exchange is possible if the essential features of the machines and their controllers are identical (DIN 1983).

For the coding of industrial robots, no standardization effort corresponding to DIN 66 025 has been undertaken as yet.

In the process chain CAD - NC programming - NC controller, the use of the above-mentioned interfaces leads to functional and interface specific information losses. Therefore at present the information flow is only in one direction and moreover redundant input and storage is necessary. Feedback is not possible either. Workshop oriented programming (manual data input) or alterations of the program at the machine lead to incompatibilities vis-a-vis the CLDATA file. During the last few years, CLDATA has not been adapted to the newly developed NC machine functions and the resulting requirements. Therefore its scope is not sufficient for modern NC machines.

3.4 Interfaces between order oriented applications and between order oriented and product oriented applications

For coupling the order oriented applications within a company with each other (for instance PPC - workshop control, manufacturing data collection - PPC) as well as for the coupling of order oriented and product oriented applications (for instance PPC - routing generation, PPC - CAD), no relevant standardization efforts have been undertaken as yet. In this context PDDI, PDES and STEP play an important role. The necessity of taking action has been expressed in the DIN reports 20 and 21 (DIN 1989, DIN 1989a). As a further step towards the definition of interfaces between workshop control and the operative machine level, MMS (Manufacturing Message Specification) with the specific companion standards can be mentioned. It is described in detail in Chapter 4 in the context of MAP.

In the area of order execution, EDIFACT (Electronic Data Interchange for Administration, Commerce, and Transport - IS 9735), though designed for data exchange between companies, presents a standardization approach. Independent of the branch, EDIFACT defines the obligatory character set, syntax of the data exchange, data segments and elements, the internationally standardized information types (for instance invoices, bills of delivery, money orders) and the telecommunication networks and services to be used. In this respect it differs from specific solutions such as SWIFT (banks), the VDA guidelines 4901 and following (automobile manufacturing), SEDAS (trade) and DAKOSY (ports). Initiated or implemented projects based on EDIFACT are ODETTE (European automobile manufacturing), COST 306 (transport), LOG (goods traffic) and EDIFICE (electronics). In the absence of standardized interfaces, many users have implemented their own individual solutions (cf. for instance Chapter 6).

3.5 Summary

For the long-term implementation of CIM concepts, open systems are an indispensable requirement. This necessitates the provision of standardized interfaces for data integration between CIM applications, as complete functional integration must remain impossible.

Interface standardization for data communication in the CIM sector focuses on the exchange of product definition and production technological data.

With regard to graphic processing on a wide range of graphic devices, GKS with its different implementation levels has assumed a most important role.

Product definition data include both geometrical and technological and administrative information. The currently existing standardized interfaces address mostly the geometry area (IGES, VDAFS, SET). Interfaces allowing the transfer of entire product models are in development (PDDI, PDES, STEP).

The single interface standards focus on certain tasks and are not suitable for the transfer of the whole range of computer internal models; their applicability is therefore restricted. Tests have shown that data losses occur during transfer or that data are being garbled; thus the standards are not yet mature. Nevertheless even today's standardized interfaces already have many technical and economic benefits. They are the result of international efforts and have an influence on the development of STEP.

Interfaces for the exchange of production technological data are applied in the automated programming of NC machines and industrial robots. They are situated between the NC (or RC) programming system and the machine itself (CLDATA, IRDATA). Their orientation is uni-directional from CAPP to CAM.

Within the process chain CAD - NC programming - NC controller, functional and interface related information losses do occur, this being one reason why information flow is currently generally uni-directional, and furthermore redundant input and storage is necessary; feedback is not possible.

Up to now no separate standardization projects have been launched with the aim of developing interfaces between order oriented applications and between order and product oriented applications. Also in this context PDDI, PDES and STEP play an important role. For the integration of order oriented applications of different companies, EDIFACT presents one standard.

The heterogeneity of CIM applications and also of their corresponding data necessitates the development of object oriented data models which are functionally independent, thereby allowing the establishment of continuous process chains. STEP is moving into this direction on the basis of an all-round product model.

The interface standardization process in the CIM area is still in its infancy. Interfaces for the transfer of complete product models which can be applied in practice will not be available for some years to come.

The few standardized interfaces available today should be applied in CIM projects in spite of their deficiencies, as they already lead to considerable technological and economic advantages and as a rule form the basis for the enhanced standards of the future.

Chapter 4

Local area networks

Satisfactory integration of all CIM related areas within a company by data technological means can be achieved neither off-line with the help of external data carriers, nor on-line with specific point-to-point connections. It must be implemented on-line, using heterogeneous networks which are independent of a specific vendor.

Due to aspects of integration, vendor specific networks will not be discussed in further detail.

Networks allow data communication between different data processing applications as discussed in Chapters 2 and 3, under conditions prevaling in real manufacturing companies.

Depending on the focus of the application, the actual requirements such a network must meet can be very diverse. As a maximum demand one could state that such a network should be able to transfer an unlimited volume of data across vast distances under real-time conditions without any faults occuring and at minimal costs. This, however, is much like having your cake and eating it. There is no such thing as 'the universal network'. In practice, networks are implemented which meet the individual demands of this maximum requirement to a certain degree. For practical reasons it is therefore more useful to apply different networks for different tasks within one company. These networks can be connected with each other.

Generally speaking, one can differentiate between wide area networks (WAN) and local area networks (LAN).

Wide area networks are public networks covering larger geographical areas such as Datex-P or Datex-L (and, in future, ISDN), which in Europe normally fall under the administration of the national postal services. They play a role in the integration of several independent companies.

Local area networks are restricted to private sites; they are usually installed within a single enterprise for handling the communication of decentralized data processing applications. They serve for the bit-serial transfer of information between decentralized systems, allow common access to locally stored data, and by providing a connection to specialized computers they enlarge the functional scope. Expensive and specialized peripheral devices are profitable only if common access is possible. Further advantages of LANs are that computers with a high work-load can be relieved by distributing tasks to other, less committed systems, and that tasks can be processed in parallel on several computers. Future applications will include computerized image and speech transfer (cf. Feiten 1990).

The technologies of LANs differ primarily with regard to topology, transfer medium, transfer method, and medium access method.

4.1 Main characteristics of LANs

The term 'topology' addresses the form of physical connection between the terminal devices linked to the network. Basically, there are three forms: star, ring, and bus. From these three main types, combinations can be implemented in inter-meshed systems.

The topology has a strong influence on performance, expansion, and fail-safety of a network.

In a star network, the data are always transferred via a central station which accomplishes the desired connection. This central station is therefore the critical point with regard to fail-safety and performance. If another station fails, the network remains operative.

In a ring network, the linked stations are connected to their respective neighbour; thus a closed ring is formed. Each station has an in and an out channel. The communication is thus performed in a predefined direction from one station to the next until the receiving station is reached. Each station has to participate in the communication, even if it is neither the original transmitter nor the final receiver of a message. If only one station fails, the whole network breaks down unless special precautions have been taken, for instance automatic short-cutting of a station which is out of operation. Enlargements of the ring can be achieved easily, as there are merely point-to-point connections between the stations. A ring can cover a large spatial area.

In a bus network, all stations are linked to a single, continuous, common transfer medium; they all have equal privilege. Communication is bi-directional. Only the transmitting and the receiving station have to be active to achieve successful communication. The breakdown of one station therefore is of no

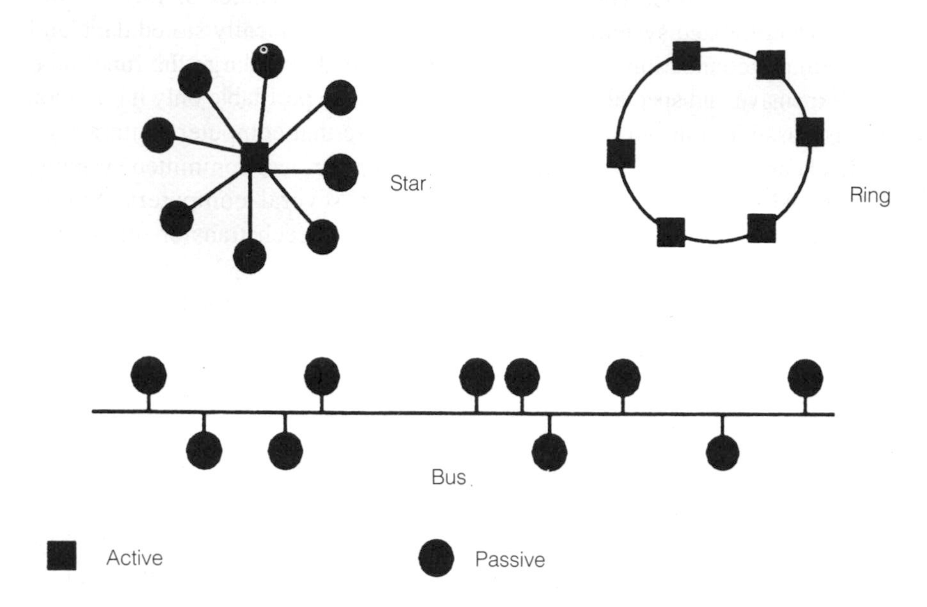

Fig. 4.1 Basic LAN topologies.

consequence to the network operation. In contrast to star and ring, the communication can be performed without delay caused by the linked stations. The number of connections to the network can easily be enlarged or reduced due to the passivity of the linked station. The size of the network is limited (limiting parameters are transfer times and reflections of the signals, the length of the bus, and the power of transmission).

Within companies the bus structure has proved to be the most efficient one.

A further characteristic of local area networks is the physical transfer medium used. Twisted two-wire or four-wire cables, coaxial cables, glass-fibre cables, and in some exceptional cases radio relay, can be used.

Cost, capacity and fail-safety are the essential criteria determining the choice of the transfer medium. Because of the already existing telephone technology, a twisted wire network covering the whole area is normally available within a company. Due to the small bandwidth, the transfer rate is not higher than 10 Mbit/sec. Moreover, this sort of cable is very susceptible to external electromagnetic interference.

Coaxial cables are less susceptible to interference and enable transfer rates of 50 or 350 Mbit/sec, depending on whether baseband or broadband transfer is used. Installation costs are higher than in the case of twisted wires.

Glass-fibre cables are wholly impervious to electromagnetic interference. The transfer rate can be as high as approximately 2 Gbit/sec. However, installation costs are still very high and some technical problems remain (passive coupling, ageing, large bending radii).

The most common type of transfer medium in LANs applied today is the coaxial cable although, especially in the manufacturing area, the use of glass-fibre cables will substantially increase.

A further important characteristic of Local Area Networks is the transfer method.

In the case of baseband transfer, the whole bandwidth of the transfer line is used, i.e. the transferred signal uses the whole transfer medium. As a consequence, only one signal can be transferred at a given time (time multiplex). The signal can be transferred either digitally or analogously as a modulation of frequences. In the case of analogous transfer, the medium is often referred to as the carrier band.

The transfer capacity is limited as only one channel with limited capacity is available. Lower costs and simpler installation form an advantage over broadband transfer.

Broadband transfer enables the simultaneous transfer of several signals. This is possible because the signals are modulated onto different carrier frequencies (frequency multiplex). Modems are necessary. Usually stations transmit on a predefined frequency which is transformed to the reception frequency of the receiving station by so-called head-end remodulators (mid-split technique). Due to the necessary amplifiers, the transfer of the signals is uni-directional. The whole frequency bandwidth of a broadband system is therefore divided into a forward frequency area and a backward frequency area. For reasons of safety, an unused buffer zone is located between them (mid-split).

The advantage of broadband systems lies in the parallel transfer of signals at substantially higher transfer rates. Moreover, broadband systems also support the parallel transfer of graphic, audio, and video signals. In comparison with baseband transfer, broadband transfer involves a higher technical effort due to the necessity of modems, head-end remodulators, as well as amplifiers and equalizers. The installation points of amplifiers and equalizers have to be carefully calculated, and the installation is costly. The locations of stations in the network which are projected for the future must be included in the planning, and they have to be simulated by resistors, otherwise the locations of amplifiers and equalizers have to be calculated again. The distance broadband systems can cover is higher than 10 kilometres and therefore exceeds the range of baseband systems.

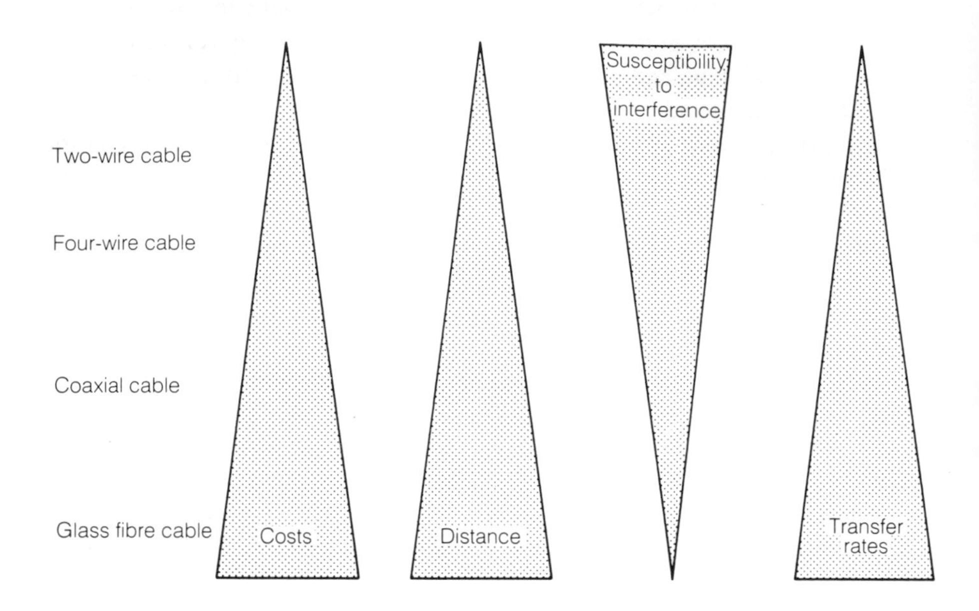

Fig. 4.2 Advantages and disadvantages of transfer media.

The last essential characteristic of local area networks is the medium access method used, which determines the access of the connected stations to the transfer medium for signal transfer.

In general, a difference can be made between a central (master/slave polling) and a decentralized control of access. The latter has a lower degree of fail-safety but is more clearly arranged. Usually, in LANs, decentralized access control is implemented.

These access methods have been standardized by the IEEE (Institute of Electrical and Electronic Engineering) as Standard 802 and have been accepted by ISO as DIS 8802. It differentiates between stochastic (non-collision-free) and deterministic (collision-free) methods of access, reservation, and the ensuing mixed forms.

In the following, the most important access methods are described: CSMA/CD (IEEE 803.3) as a stochastic method and the token access method (ring: IEEE 802.5, bus: IEEE 802.4) as a deterministic access method.

CSMA/CD (carrier sense multiple access / collision detection) is also known as the ETHERNET access method and has been developed for bus topologies. A station intending to transmit scans the medium to see whether it is free (carrier sense), and transmits if the medium is not in use. Collisions may occur when several stations simultaneously start to transmit because the scanned medium was detected as free by all other scanning stations, too (multiple access).

In order to detect such collisions (collision detection), all transmitting stations are obliged to continuously scan the medium while transmitting. If several stations are transmitting simultaneously, they detect the external signals and must interrupt their transmission. Only after a random waiting period and only if the medium is free may they start a new attempt to transmit.

In case of a high work-load being put on the network, i.e. if many stations simultaneously try to access the network, frequent collisions are inevitable and network access is paralysed. An upper time limit for successful access to the transfer medium cannot be determined.

In this respect, the token access method is physically different: token bus (IEEE 802.4), token ring (IEEE 802.5).

Here a token is represented by a predetermined bit pattern circulating on a logical ring between the connected stations. The physical topology of the network is of no consequence. When a station receives a token, it transmits the token to its logical neighbour if it does not intend to send. If, however, it intends to send, it changes the token's bit pattern from 'free' to 'occupied', first transmits the

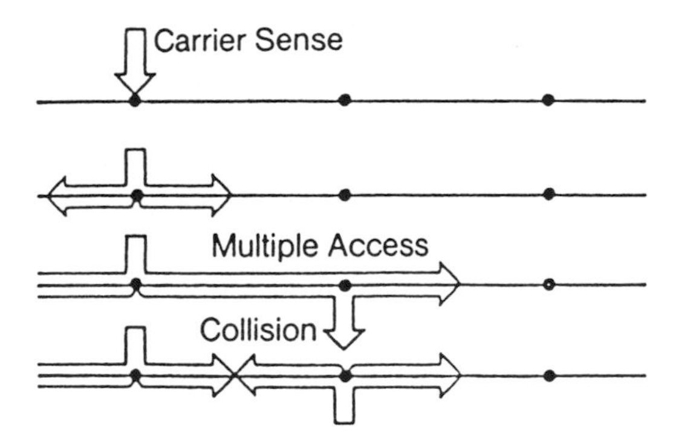

Fig. 4.3 CSMA/CD method (Wolf 1986).

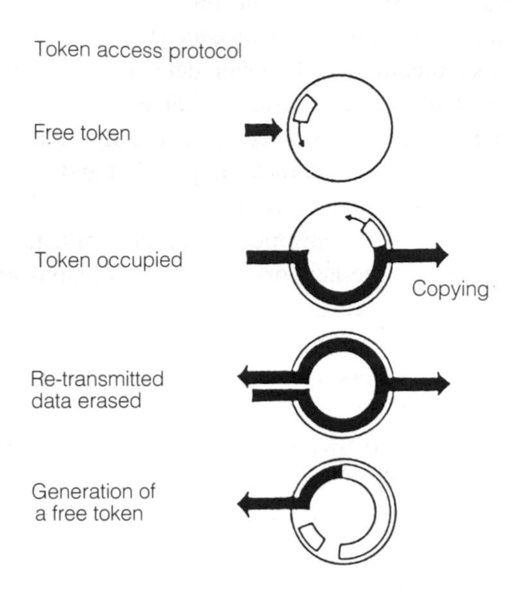

Fig. 4.4 Token ring method (Wolf 1986).

changed token and then one or several messages (the length of which is normally determined by a time limit). The transmitting station keeps a copy of the messages it has sent. The logically adjacent stations check whether they are intended to receive the message. If this is not the case, they send the message to their logical neighbour. Otherwise the receiving station copies the message and transmits the original message together with a receipt back to the sender. When this station receives the receipt, and if the stored copy is identical with the original message on the medium, it erases the message from the logical ring and generates a free token, which it then sends to its logical neighbour.

It is possible to differantiate between single and multiple token methods.

In contrast to CSMA/CD, for each station a maximum amount of time until the next medium access is deterministically established (number of stations multiplied by maximum transmission time). This can be especially important in the case of manufacturing applications. A high work-load does not result in paralysis of the network. (For more details on network characteristics see for instance Kauffels 1985, Suppan-Borowka *et al.* 1986, Eckardt *et al.* 1988, Madron 1988, Martin *et al.* 1989, and Pimentel 1990).

4.2 ISO/OSI reference model

The reference model of the International Standards Organization (ISO) for communication within open systems, i.e. Open System Interconnection (OSI), was established in 1983 (ISO DIS 7498). It has a fundamental relevance for the structuring of complex communication procedures between data processing systems, and basically not only for communication within open systems, but also within closed systems. In closed systems, vendor specific communication protocols (e.g. SNA (IBM), DECNET (DEC), TRANSDATA (Siemens)) are used. These systems are in principle based on models similar to the ISO/OSI reference model, but differ in details which result in incompatibility. In contrast to this, open systems are based on the ISO/OSI reference model which enables open communication of data processing systems of different types without the need for special adaptations.

The reference model merely describes the communication between open systems but not their internal functionality. Neither is it a specification for implementation, nor does it determine the technology of the systems or the transfer media. Thus, this model is a standard which contains functional provisions for further, more specific standards.

The ISO/OSI reference model subdivides the complex communication procedure into partial steps. A seven layer model was developed which in a hierarchical manner attributes all the functions occuring in a communication procedure to a corresponding special layer.

Before turning to the significance of the single layers, some of the basic features of the model will be discussed.

Each layer, with the exception of the top one, provides certain defined services for the layer above it. Each layer, with the exception of the bottom one, uses the services provided by the layer below it. The interfaces between the layers are determined on an abstract level (vertical protocols). The single layers have to be sequentially passed through during a communication.

Also within the layers of the communicating systems, communication protocols (peer-to-peer-protocols) are defined (horizontal protocols). Each entity of the layer N of a system communicates exclusively with the entities of the layer N of other systems. This communication between layers of the same level is implemented by each layer adding to the transferred user data further specific information and additional control information. This procedure is controlled by the protocols.

Depending on the application, some of the layers do not have to be implemented, and may be empty. This, however, may result in incompatibilities.

Generally speaking, the layers of the ISO/OSI reference model can be put into two categories.

The transport system (layers 1 - 4) transports user data without regard to their syntax or semantics and without influencing the processing of the transferred data. After sucessful transport of the user data by the transport system, the attribution and the processing of the user data is controlled by the application system (layers 5 - 7).

In a final system, all seven layers of the ISO/OSI reference model are implemented, while in a transit system only the layers 1 - 3 are implemented.

The lowest layer 1, the physical layer, determines the mechanical, electrical and electromagnetic coupling of the systems to the transfer line. This includes transfer media such as cables and plugs, as well as the electrical interface characteristics such as data transfer rate and the electrical representation of the signals. Examples for defined standards are X.21, X.24, V.24, and V.28.

IEEE 802 determines the transfer technologies and connection methods for LANs (see Section 4.1).

The data link layer (layer 2) has the task of detecting possible transfer errors which have occured in the physical layer and - if possible - to correct them. This is done by checking the message with added redundancy information (added check sums). In case of errors, the receiver effects a new transfer by sending a message to the transmitter.

A further task of this layer is flow control, i.e. the synchronization of the transfer speed of one system with the reception speed of another, and attributing sender and receiver addresses to a message. This layer also conducts and controls the medium access (see Section 4.1, medium access methods).

The network layer (layer 3) is mainly relevant for WANs or for the transfer of messages via several local sub-networks. Within one local network, this layer may be empty, because in this case a direct connection between two systems is usually effected by the data link layer. Otherwise the network layer defines the optimal way of transferring messages throughout the whole network. Optimization criteria can be transfer costs, times, data rates, or data security.

The network layer on the one hand enables a link oriented transfer of messages, i.e. that all messages between two systems follow the same path. On the other hand the network layer can transfer the messages without a link, i.e. the messages between two systems are given an address and can pass through the network to the receiving system on different paths. In the latter case it may happen that one message overtakes another, with the effect that the sequence of messages at the sending and the receiving stations does not coincide (the correct sequence of messages is then restored by layer 4). A further important task of this layer is the multiplexing of connections passing through the same network area. This includes the linking of messages to packages in order to achieve a more efficient use of the

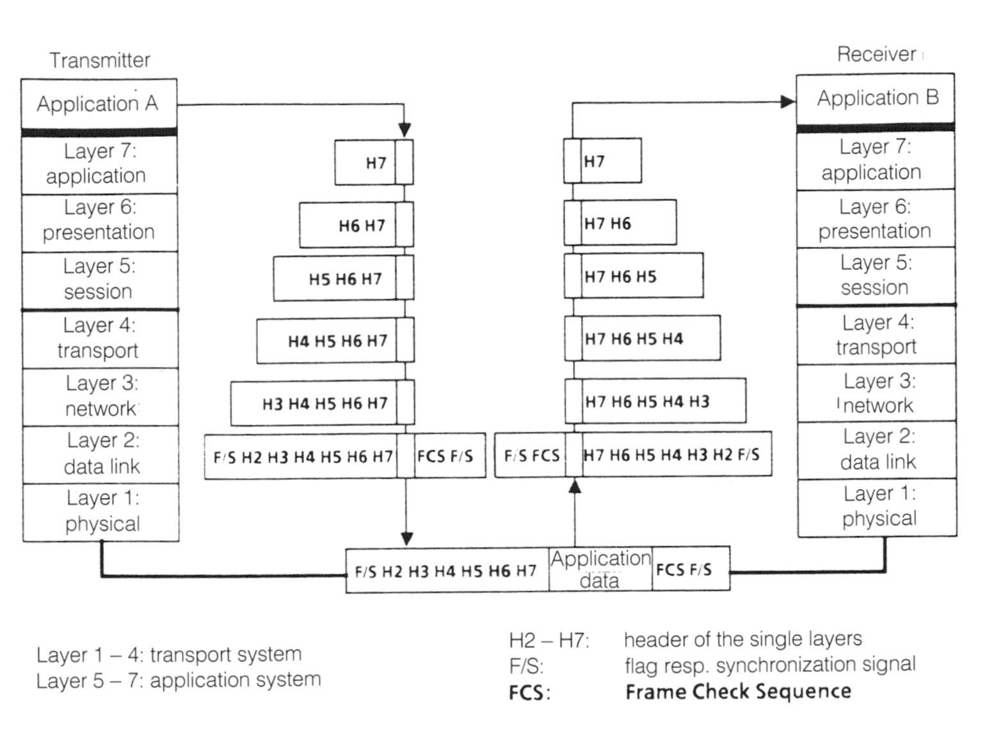

Fig. 4.5 Connection structure between two systems according to the ISO/OSI reference model (with reference to ZVEI 1987).

network and splitting them up again after transmission, as well as the transmission of so-called priority data which are to be sent with a higher priority than the normal messages.

The transport layer (layer 4) is the border layer between transport and application system. The network layer supports the connection between the computer systems. The transport layer goes even further and effects an end-to-end connection between the processes running on the computer systems. It is therefore the task of this layer to separate transport related aspects of the message transfer from application related aspects. Primarily, the messages sent by the network layer which have a limited length must be transformed into messages of undefined length for the superceding session layer.

The session layer (layer 5) is a layer, the services of which must be completely conducted by the end system of a communication. It is the first layer which does not belong to the transport system, but is part of the application system.

Consequently, the session layer provides the services for the actual communication control between the end systems. This includes the organization of a communication relationship (session), the determination of the dialogue procedure, the subdivision of a communication into logic sections, reestablishment in case of errors, and finally the closure of a session.

The presentation layer (layer 6) has mainly the task of adapting the possibly differing bit representations of data (e.g. ASCII, EBCDIC) of the end systems to each other. To achieve this, the presentation layer conducts a syntax transformation (concerning data formats, definition of fixed point and floating point numbers, control characters, data coding) from a local syntax to a concrete transfer syntax or vice versa.

The application layer as the highest layer is the interface to the user, i.e. to a certain application.

Application-specific aspects are related to areas close to the operating system as well as to the specific contents of the applications.

A standardization of the latter area is difficult to achieve due to the heterogeneity of data processing applications. A classification of applications and their aspects may in time lead to the recognition of similar communication behaviours and thus open possibilities for standardization. Examples of this are interfaces for the exchange of product definition data mentioned in Chapter 3.

In the area close to the operating system, the application layer does already supply several common application programs for data communication such as file transfer, electronic mail, remote job entry or remote data access.

Many of the currently supplied networks only cover the implementation of layers 1 and 2. This is unsatisfactory in the sense that a communication acceptable for the user is only possible when at least the layers 3 and 4 have also been implemented, as otherwise users have to put much effort into writing their own individual programs. This should be taken into account when opting for a LAN.

Complying with the ISO/OSI reference model has the advantage that applications of various types can communicate with each other. However, the necessity that a communication has to pass through all layers of the reference model, i.e. because of the protocol overhead, often slows down the communication. Therefore real-time conditions as they are often given in the manufacturing area are impossible to meet. As a consequence, in a number of cases it is advisable to reduce the seven layer structure.

4.3 Means of network connections

As mentioned in the introduction to this chapter, the requirements for networks can be very different depending on the tasks they have to fulfil; therefore the application of different partial networks within one company is useful. Communication across the boundaries of such partial networks is equally beneficial. By connecting the partial networks to a whole network, company-wide communication becomes possible.

For this purpose, four connecting elements are available, the application of which depends on the compatibility of the protocols used on the single layers of the partial networks which are to be connected.

The four elements are repeater, bridge, router, and gateway.

Repeaters are mainly applied to extend the physical working area of a network, i.e. they connect similar network segments on the physical layer with each other.

A bridge links two partial networks which can be different up to the second layer. The partial networks are connected on the data link layer, i.e. the medium access methods of both partial networks may be different. It is therefore possible for instance to connect a network which uses CSMA/CD as the access method with a network which uses the token method. The sole pre-condition is that the bridge is able to store messages.

Routers are applied for the connection of networks on the third layer of the ISO/OSI reference model. All the layers of the partial networks below layer 3 may have different protocols. In contrast to bridges, no determined link address within the whole connected network is required. Consequently, a higher effort must be made on the network layer (consideration of alternative paths, address management, format adaptation, flow control).

Gateways can mainly be applied for connecting networks which follow the ISO/OSI reference model with closed, vendor specific networks. The layer on which the connection is performed is the lowest layer with common protocols of the connected networks. In the case of wholly incompatible networks, this would be the application layer (layer 7).

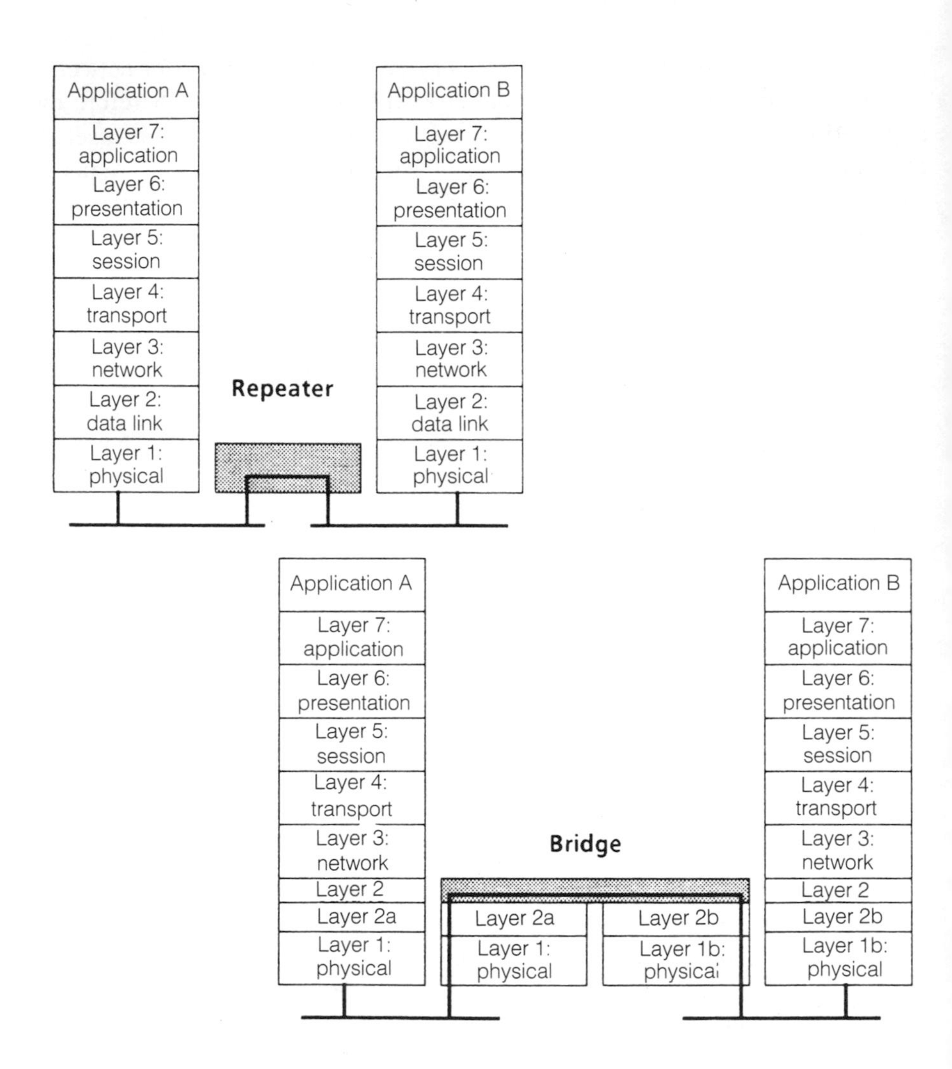

Fig. 4.6 Network connection by repeater and bridge.

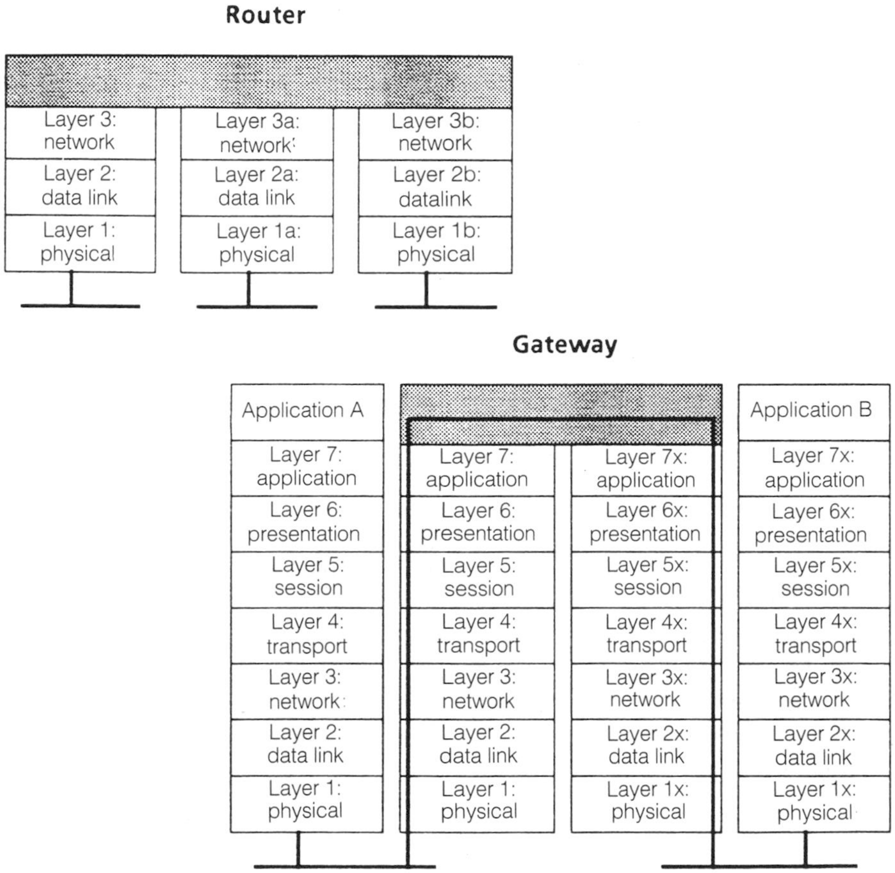

Router

Layer 3: network | Layer 3a: network | Layer 3b: network
Layer 2: data link | Layer 2a: data link | Layer 2b: datalink
Layer 1: physical | Layer 1a: physical | Layer 1b: physical

Gateway

Application A | | | Application B
Layer 7: application | Layer 7: application | Layer 7x: application | Layer 7x: application
Layer 6: presentation | Layer 6: presentation | Layer 6x: presentation | Layer 6x: presentation
Layer 5: session | Layer 5: session | Layer 5x: session | Layer 5x: session
Layer 4: transport | Layer 4: transport | Layer 4x: transport | Layer 4x: transport
Layer 3: network | Layer 3: network | Layer 3x: network | Layer 3x: network
Layer 2: data link | Layer 2: data link | Layer 2x: data link | Layer 2x: data link
Layer 1: physical | Layer 1: physical | Layer 1x: physical | Layer 1x: physical

Fig. 4.7 Network connection by router and gateway.

4.4 Heterogeneous local area networks

As already mentioned in the introduction to this chapter, the requirements for the communication can differ strongly depending on the application focus of the communicating partners. Therefore in practice within one company different networks are used for different application areas, which are connected by certain network connection means.

When analyzing the global information and communication structure of a company, hierarchies can be identified to which CIM components or their partial areas can be attributed.

For a classification of the application areas of local networks, a hierarchical subdivision of the CIM communication structure is helpful. This is briefly explained in the following paragraphs.

In Fig. 4.8, the CIM hierarchy consists of three levels. A subdivision into further levels is possible. A superior planning system (here the planning level) takes over the rough planning with the help of compressed data. The control systems of the control level have to perform the work distribution and the supervision of the work progress. The actual processing systems are located on the process level. Each system on each level may in turn have its own hierarchy of levels.

The overall system consists of a number of control loops. The control of a control loop is performed with the help of nominal values supplied by the respective superior level. When exceptional cases occur which the control loop cannot handle independently, an error message is dispatched to the superior level. In just the same way success messages and reports are given to the superior level. According to the characteristics of a company, communication relationships may exist between the control loops on one level of a hierarchy (e.g. on the process level in the case of the KANBAN system). Such information exchange between different control loops on one level or the skipping of a level to superior or inferior levels may only occur in special, defined cases.

Generally speaking, a differentiation can be made between the more important vertical communication and horizontal communication. Vertical and horizontal communication are not only characterized by data or information flow, but also by the storage of information on the levels. As a consequence of their different tasks, the different levels have different planning and processing perspectives and cycles (from top to bottom ranging from year to ms).

Data processing and communication within the system hierarchy is therefore characterized by the following facts: the periods of time decrease from top to bottom (from year to ms), and likewise the data volumes of single process and communication procedures decrease (Mbyte to bit). The number of communication processes, however, is much higher on the lower levels, and transfer times can range between milliseconds on the lower level and hours on the upper level.

The model of the system hierarchy described above plays a role in the application of data processing in hardware hierarchies, where certain tasks are assigned to single systems of varying intelligence, which correlate to the hierarchies of the model. As can be deduced from the explanations above, the intelligent systems are connected among each other by different networks. Figures 4.9 to 4.11 show different characteristics of hardware structures in medium-sized tool construction companies. In the company shown in Fig. 4.9, apart from a stock management system only point-to-point connections are implemented. (On the subject of communication hierarchies cf. Steusloff 1987, Steusloff 1989, Duelen *et al.* 1986, Albus *et al.* 1981, McLean *et al.* 1983, CAM-I 1983, and Jones *et al.* 1990.)

In accordance with the different requirements within the system hierarchy, standardization efforts for heterogeneous local area networks were undertaken from different perspectives which aim at supporting the communication of open systems at least for certain application areas: TOP, MAP, and fieldbus.

4.4.1 MAP (Manufacturing Automation Protocols)

MAP was the first among the above-mentioned standardization attempts. In 1980, the Manufacturing Automation Protocols Task Force was founded at General Motors (GM). The aim of this unprecedented initiative coming from the user side was to develop an open, heterogeneous factory communication network based on an open heterogeneous network, to which the miscellaneous systems participating in the manufacturing process, i.e. a variety of computers, terminals, controllers, robots, and CNC machines, can be connected.

A study by GM had shown that costs for communication between intelligent systems involved in manufacturing could be as high as 50% of the costs of the intelligent systems themselves if the adaptation of two systems to each other were conducted case-by-case. Given a fivefold increase in the number of intelligent systems involved in manufacturing to 200,000 in 1990, this would have led to an enormous rise in costs.

The primary aim of the MAP initiative was to achieve a reduction of costs by unifying communication. This was possible either by creating a new system specially adapted to the GM requirements, or by taking over and further developing already existing standards.

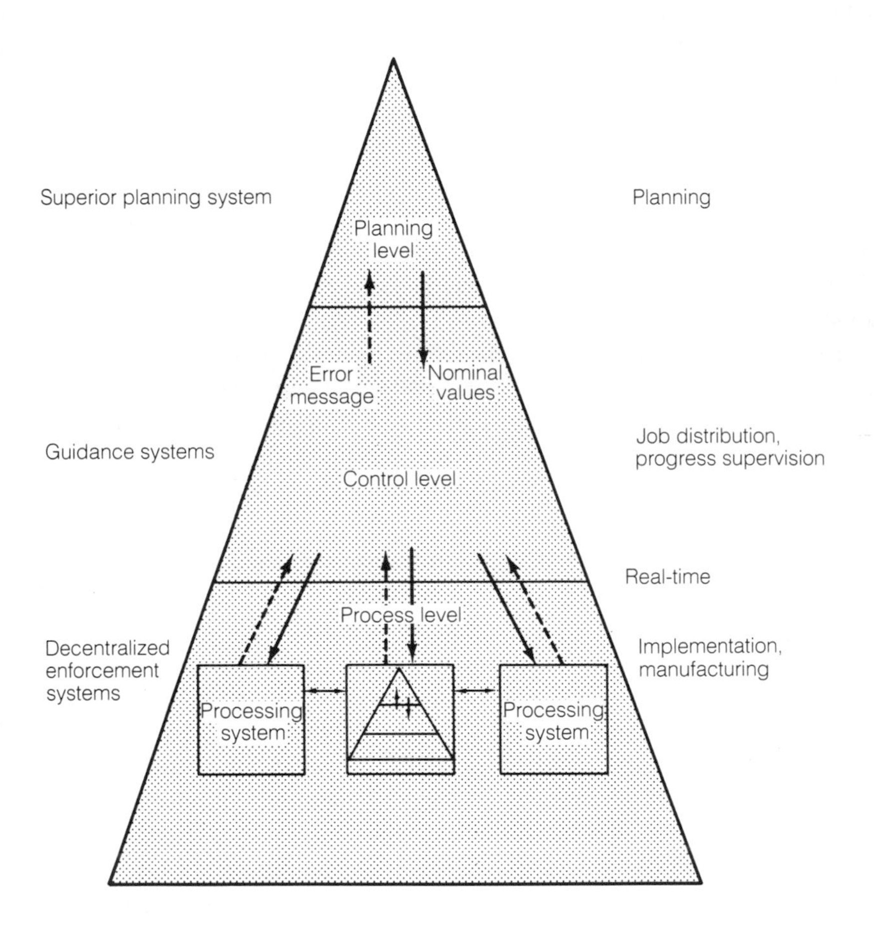

Fig. 4.8 Model of a system hierarchy.

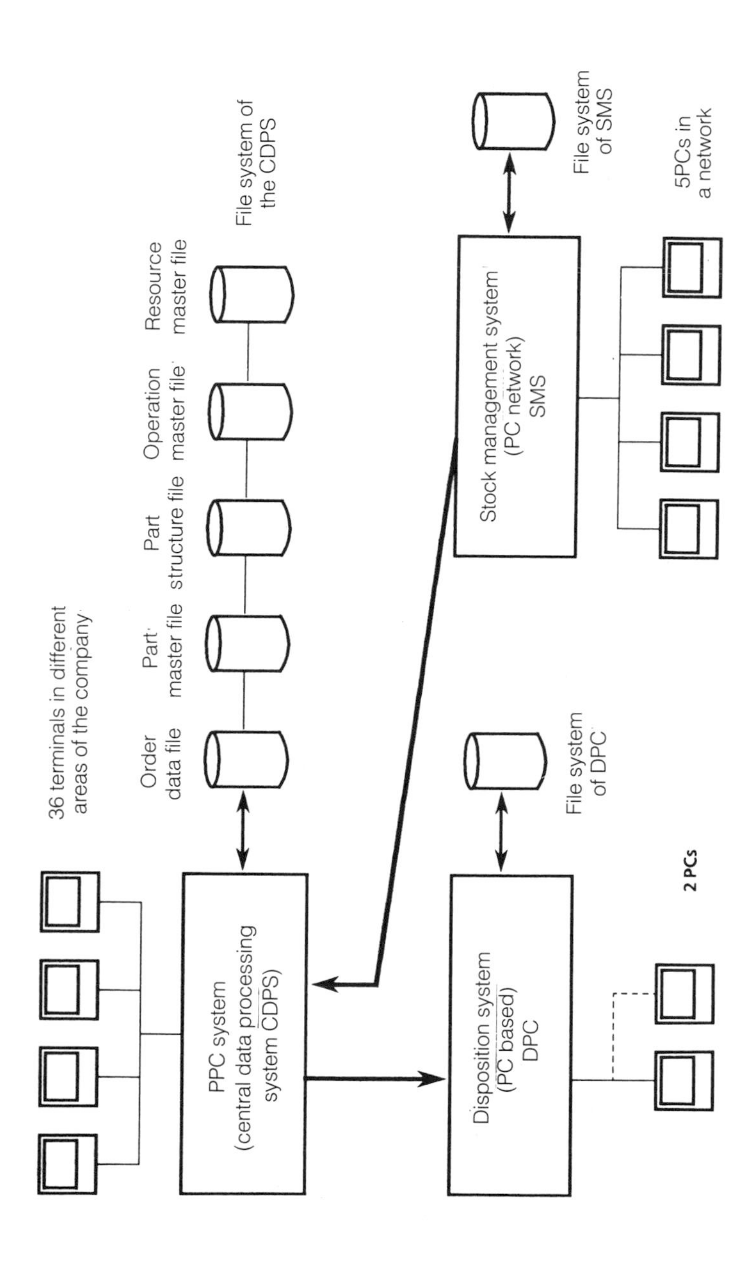

Fig. 4.9 Hardware configuration and network structure: company A.

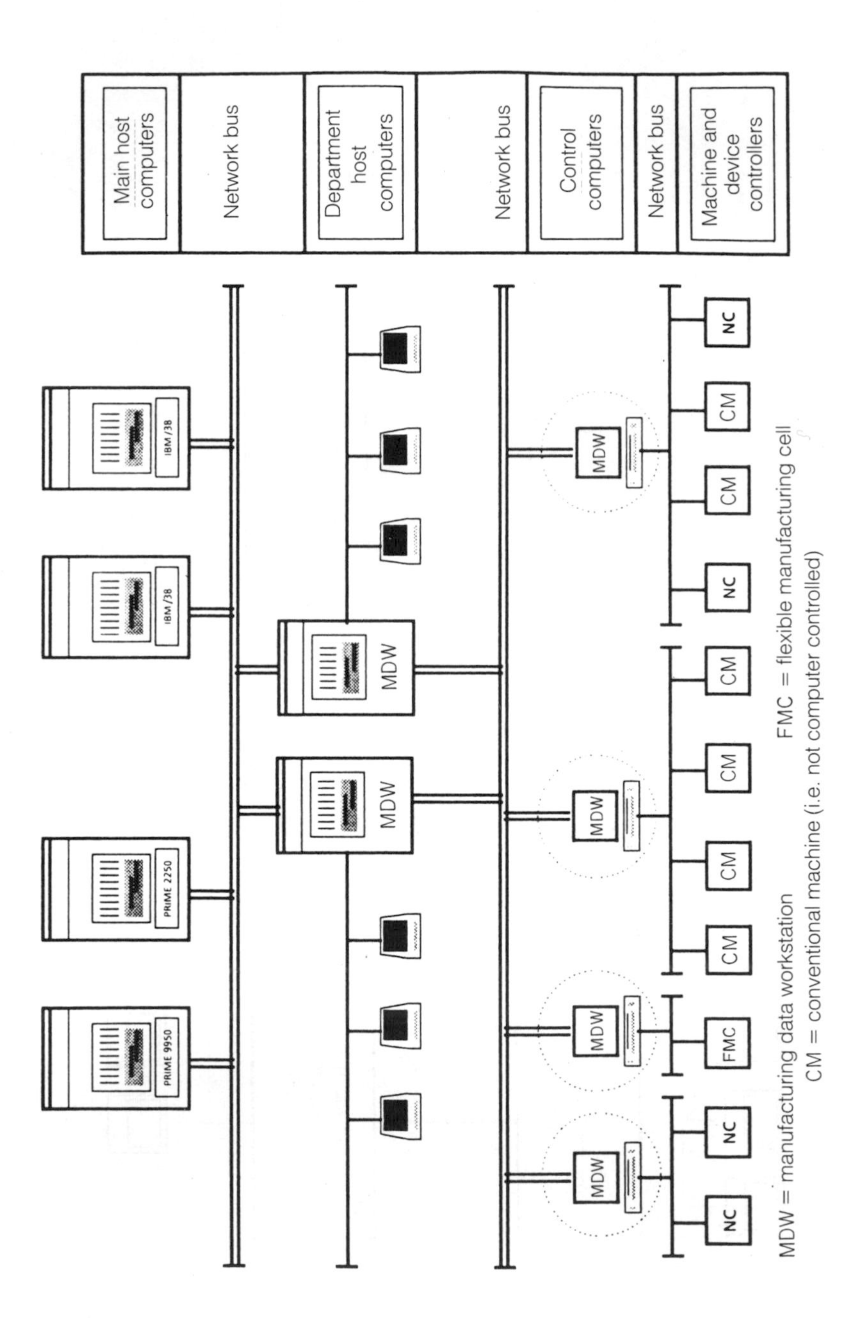

Fig. 4.10 Hardware configuration and network structure: company B.

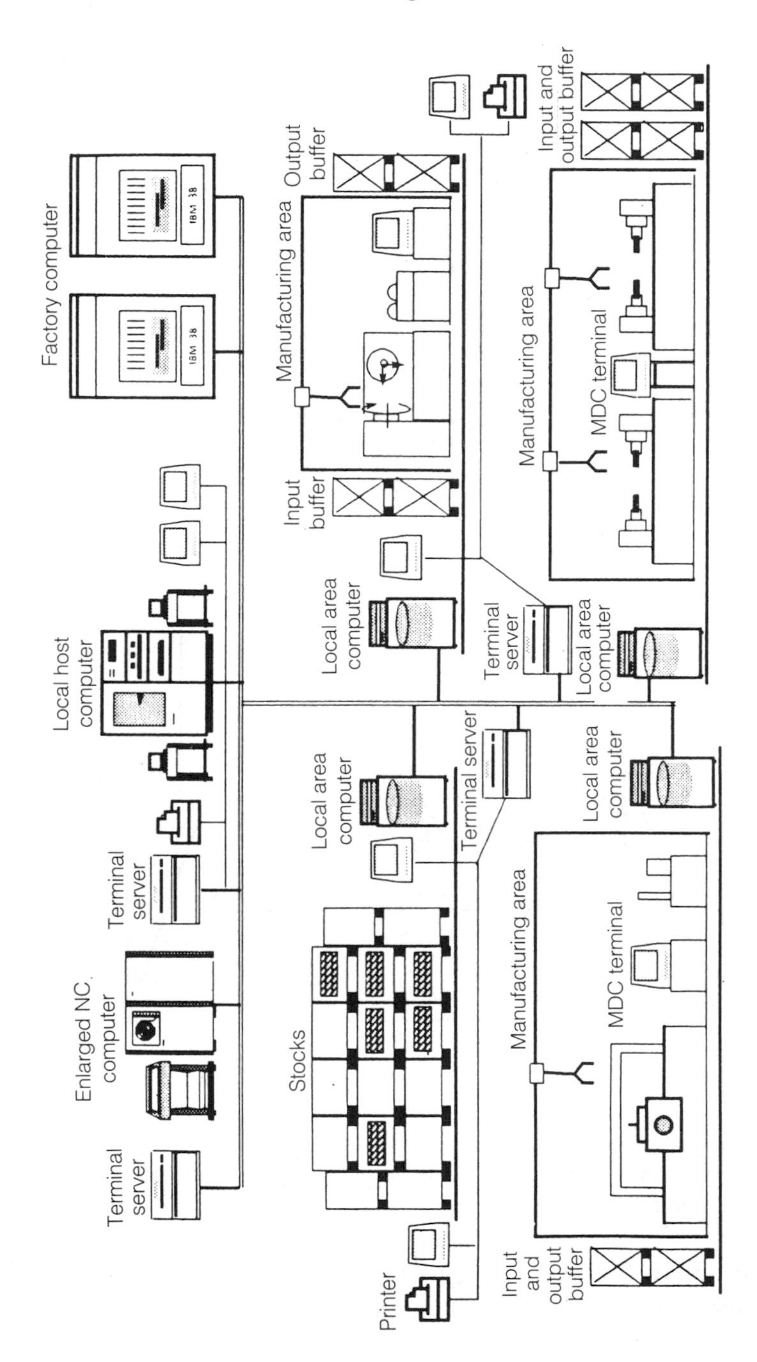

Fig. 4.11 Hardware configuration and network structure: company C.

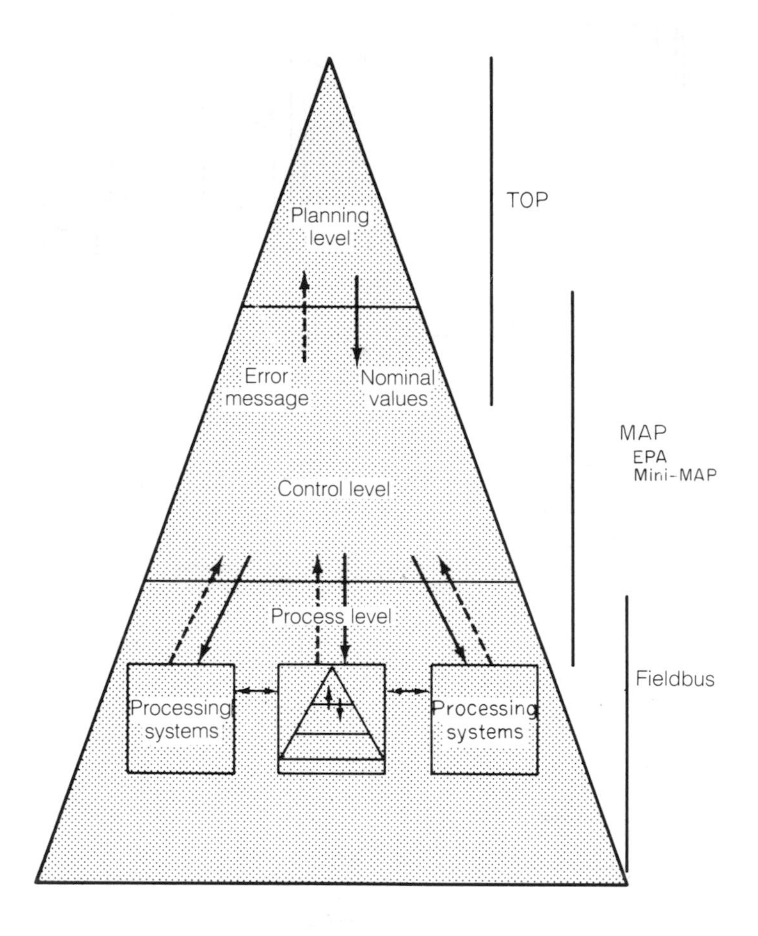

Fig. 4.12 Attribution of standardization approaches for networks to the system hierarchy.

As a basis for MAP, GM chose the ISO/OSI reference model which is to be filled with existing and expected standards in order to adapt it to the MAP requirements.

Because of GM's strong influence as a purchaser, the cooperation of a large number of system vendors of the greatest variety could be secured. Furthermore, cooperation could be achieved with important industrial partners such as Ford and General Electric.

Since the foundation of the Manufacturing Automation Protocols Task Force in July 1980, the following milestones of the MAP development have been reached:

	1982	Foundation of a working group of GM together with IBM, DEC, HP and others.
	1984	Foundation of the MAP Users Group (MUG).
	1984	Installation of a demonstrator at the National Computer Conference in Las Vegas (Nevada) including computer systems of seven vendors: MAP 1.0.
February	1985	MAP 2.0.
May	1985	MAP 2.1.
August	1985	MAP pilot installation at GM in Marion (Indiana).
November	1985	Demonstrator installation at the Autofact '85 in Detroit (MAP 2.1) with 21 hardware vendors.
November	1985	Foundation of the European MAP Users Group (EMUG) comprising 50 users and 50 vendors; more than 20 vendors declare their intention of marketing MAP compliant products.
August	1986	MAP 2.2.
June	1988	Fair installation at the Enterprise Network Event in Baltimore on the basis of MAP 3.0.
October	1988	Fair installation in Europe at the SYSTEC '88 in Munich, organized by EMUG.
	1988	Stability of MAP 3.0 specification guaranteed for six years.

MAP version 3.0, which is currently the one to be implemented in factory environments, is described below. The preceding versions are not upwardly compatible with MAP 3.0.

Layer 1

As a physical medium a coaxial cable with broadband transfer is recommended. The transfer rate is 10Mbit/sec. The choice of coaxial cables resulted from its relatively widespread availability in the U.S. It was formerly used for cable television (CATV in the 1950s) and had already been installed in many factories for use in LANs, also at GM.

Because MAP uses only three channels with a bandwidth of 12 MHz each, further signals such as speech and video can be transmitted simultaneously, and several networks or partial networks can operate with the same cable. With a broadband system, relatively large distances - up to 10 km - can be covered.

As a low-cost solution for smaller installations with up to 32 stations, a carrier band with a transfer rate of 5 Mbit/sec can be used. In this case, the maximum distance is 700 m.

Layer 2

As MAP 3.0 is based on the IEEE 802 protocols, the data link layer is subdivided into two layers. The LLC sublayer (logical link control) uses the connectionless service (ISO 8802/2). For the MAC sublayer (medium access control), the token bus method (ISO 8802/4) was chosen as the medium access method because, in the factory environment for many application, defined transfer times must be guaranteed. Due to the passive coupling to the network, the breakdown of a single station does not affect the network.

Layer 3

Also on the network layer, MAP 3.0 uses a connectionless service (CLNS - Connectionless Mode Network Service); therefore also on this layer flow control, segmentation, and sequence maintenance cannot be performed.

Layer 4

The transport layer has to provide the connection orientation with flow control, segmentation, and sequence maintenance, apart from its substantial task of error detection and recovery.

Layer 4, being the highest layer of the transport system, provides a unified service interface independent of the implementation of the network technology for layer 5, which is the lowest layer of the application system. Therefore protocols of the lower layers can be exchanged or modified without affecting the application system.

Layer 5

The session layer is the lowest layer providing functions for communication management and synchronization between the applications. MAP 3.0 specifies full-duplex connections together with the employment of the session kernel as a subset of ISO 8327.

Table 4.1 Specification of the ISO/OSI layers in MAP 3.0 (bold characters: identical with TOP 3.0)

Layer	MAP 3.0 specification
	Application Program Interfaces (API)
Layer 7: application	ISO-MMS DP 9506 **ISO-FTAM DIS 8571** **ISO-ASCE DIS 8649/1-2** **DIS 8650/1-2** **MAP-Directory Services** **MAP-Network Management**
Layer 6: presentation	**ISO-Presentation DIS 8823**
Layer 5: session	**ISO-Session Kernel IS 8327 duplex**
Layer 4: transport	**ISO-Transport Class 4 IS 8073**
Layer 3: network	**ISO-Connectionless Internet DIS 8473**
Layer 2: data link	**ISO Logical Link Control DIS 8802/2** Type 1 (for broadband) Type 3 (for carrier band) - - - - - - - - - - - - - - - ISO- Token Bus DIS 8802/4
Layer 1: physical	ISO Token Passing Bus DIS 8802/4 Type 1 (for broadband) (or) Type 3 (for carrier band)

Layer 6

The presentation layer provides functions for coding and transformation. Its purpose is the solution and the management of syntax differences of the data which are exchanged between two applications. MAP 3.0 uses the functionality of the presentation kernel functional unit as a subset of ISO DIS 8823.

Layer 7

In order to be able to serve different applications, MAP 3.0 specifies different protocols in layer 7.

For the communication of two programs on different computers, ACSE (association control service element) provides the services of establishing and terminating a connection and transmitting and receiving data records.

FTAM (file transfer access and management) supplies services for data transfer and generating and erasing data files on computers.

MMS (Manufacturing Message Specification - ISO 9506) is specially suited for the communication with intelligent devices in the manufacturing area such as CNC machines, robots, and others. MMS has been standardized as RS-511 by the Electronic Industry Association (EIA).

Among others, MMS provides the following functions important for distributed systems:

- variable access;
- message passing;
- resource sharing (synchronization);
- program management;
- event management (Pimentel 1990).

To give an example, within program management, instructions e.g. for CNC machines, such as activating or terminating a machine cycle (start cycle, end cycle), identification of parts (part), axis manipulations (axis offset) or manipulation of part palettes (exchange), are defined in standardized formats.

Therefore MMS plays a major role in achieving the purpose of MAP, i.e. the implementation of a unified communication infrastructure within a factory.

The MMS structure follows the client-server principle. A client, i.e. an intelligent device, sends a request as a message to a server - also an intelligent device - which then executes the order and sends back a message to the client indicating the results. MMS in an object oriented form describes what a server can understand and perform. A server, itself an object, is defined as a so-called virtual manufacturing device (VMD), i.e. only those functions and characteristics of a real device which are open to the outside are represented. The internal operations of the real device are irrelevant. MMS contains 16 manufacturing relevant object types and 79 operations which can be performed on the objects. Representation of the objects and operations, and in exceptional cases also enlargements for certain types of devices (PLC, RC, NC, process control, etc.) are possible or have already been defined. They are the so-called companion standards (Schümmer 1988, Schwarz 1989, Pimentel 1990).

(To meet the requirements of the processing industries, EIA is working on a PMS (Processing Message Specification)).

MAP provides user interfaces which make the application programs independent of the implementation details of the network services (API - Application Program Interfaces).

The basic concept of MAP calls for a background network (backbone) which links the isolated automation solutions or provides a connection to superior control levels. The second important factor of this concept is the so-called cellular concept. Cells are usually single sub-networks with protocol architectures which do not necessarily comply with MAP. Routers are applied for network connections between the MAP backbone and the cells arranged in MAP. Connections with other, homogeneous networks are normally achieved with gateways.

The necessity of sub-networks arises from the conceptual characteristics of the backbone network or from specific requirements on the cellular level. These requirements include for instance the necessity of real-time data processing which cannot be achieved by the backbone network. This is primarily a consequence of the high functionality of the seven layer protocol architecture, the sequential execution of which, for instance in the case of a file transfer, may take several seconds, even if the files are very small. A further disadvantage with regard to response time is the necessary technical effort for the broadband systems as well as the vast space a network may cover and the large number of stations connected to it. Furthermore, an implementation of all seven layers for simple and therefore inexpensive intelligent devices which are to be connected to the MAP network, such as for instance programmable logic controllers (PLC), would lead to unacceptable costs.

For these reasons a reduced architecture for cells was defined, the so-called Enhanced Performance Architecture (EPA).

Here the layers 3 to 6 of the ISO/OSI reference model are empty. In contrast to the normal backbone, a baseband with a token bus medium access method (carrier band) and a data transfer rate of 5 Mbit/sec is used. The LLC protocol (class 3) is enlarged. It optimally supports the return of direct receipts or answers to the sender. Layer 7 is filled with a reduced MMS version, the so-called Mini-MAP MMS. Alternatively, it can be replaced by a Mini-MAP object dictionary or a user defined protocol. It is possible for EPA stations to communicate directly with stations at the backbone via layer 7. Because of this, ISO/OSI compatibility can be achieved for certain application cases which are covered by Mini-MAP MMS.

For the structuring of EPA, the requirements put forward by the processing industry which have been defined in the PROWAY-C specification play an especially important role.

A further alternative with reduced architecture is Mini-MAP. It differs from EPA in the fact that layer 7 is also unspecified. Therefore ISO/OSI compatibility is not possible. A direct communication of Mini-MAP stations with stations at the backbone is thus impossible, too. It can only be done via a so-called cell-controller which takes over the gateway functions.

The cell-controller also performs further specific functions for communication within a cell which go beyond the control functions of the two lowest ISO/OSI layers.

The MAP backbone is especially suitable for connecting computers on the control level to a network, such as factory, workshop and cell computers for the non-time-critical transfer of files and NC-programs. With the aid of sub-networks such as EPA, Mini-MAP or homogeneous networks, intelligent units can be arranged in a network on the process level. This assures real-time performance for critical applications and lower connection costs due to the less complex protocol requirements of the single units. Fail-safety is considerably enhanced by the cellular concept. The work-load on the backbone is reduced by local communication within the cell.

In order to assure the compliance of products with the MAP specifications, test centres were established by NIST in Ann Arbor in the USA and in Europe for example by the Fraunhofer Institute for Information and Data Processing (IITB) in Karlsruhe, Germany. They supply certificates according to unified criteria.

In Europe, within the framework of the ESPRIT program, five vendors (Bull, GEC, Olivetti, Siemens-Nixdorf, and TITN) and four users (Aeritalia, British Aerospace, BMW, and Peugeot), are jointly working on the development of MAP products in the project CNMA (Communications Network for Manufacturing Applications). Their experiences have an influence on MAP development as well as standardization efforts in this area in general.

Generally speaking, the evaluation of the application prospects for MAP should not be too optimistic. Products for hardware integration which comply with MAP can already be bought on the market. However, as mentioned above, the standardization protocols on some of the layers are not yet mature. Even some already standardized ISO protocols contain substantial faults or are not yet unequivocally specified.

MAP products used for test or pilot installations are mostly not yet cleared for marketing. Available MAP products often show incompatibilities due to an

Table 4.2 Specification of the ISO/OSI layers in EPA

Layer	EPA specification
Layer 7: application	MINI-MAP MMS or MINI-MAP Object Dictionary oder User defined
Layer 6: presentation	NULL
Layer 5: session	NULL
Layer 4: transport	NULL
Layer 3: network	NULL
Layer 2: data link	ISO-Logical Link Control DIS 8802/2 Type 3 ISO-Token Bus DIS 8802/4
Layer 1: physical	ISO-Token Passing Bus DIS 8802/4 Carrierband (5 Mbit/sec)

incomplete or not yet unequivocal standardization. Prices for direct connections to the MAP network currently range between 5,000 and 10,000 DM. The costs will only decrease once VLSI hardware is commonly available.

For the time being, the installation effort remains relatively high and requires high qualifications on the part of the installation personnel.

Today, MAP products merely allow the building of test installations and therefore the gathering of know-how and experiences. With an eye on a long-term scheme in the framework of strategic planning, organizational planning and the construction of partial networks based as far as possible on MAP compatible products should be started now. The investments can be secured by a later integration into a MAP backbone.

(Further literature: Bauer 1987, CIM 1986, Kommtech 1986, Suppan-Borowka 1986, Nagakawa *et al.* 1988, Cooling *et al.* 1989, Nakano *et al.* 1990, Roach 1990, and Pimentel 1990.)

4.4.2 TOP (Technical and Office Protocols)

In parallel with the works on MAP conducted at GM, the development of TOP (Technical and Office Protocols) was undertaken under the guidance of Boeing Computer Services.

When viewing the system hierarchy and the position of MAP in it (see Fig. 4.12), TOP forms an upward enlargement of the application area of MAP.

In general terms, the aim of TOP is the development of an open, heterogeneous communcation system which connects the various applications in the design and administration areas (office). Due to the increase in performance of microprocessor computers, such applications have been installed in an increasingly decentralized manner on heterogeneous office workstations.

Because of the organizational process structure, it is useful to connect such heterogeneous and isolated data processing areas with each other (the paperless office).

The main requirement is to allow the exchange of structured and unstructured data volumes of graphic documents and documents with mixed graphic and textual elements, and the access to central transaction systems of functional type as well as to central database systems.

As with MAP, the basic idea behind the development of the concept was to make the best use of existing international standards and the development or enlargement of standards. TOP is also based on the ISO/OSI reference model although it is filled with protocols suitable for applications in the area of office communication.

At Autofact 1985 in Detroit, TOP was demonstrated together with MAP for the first time. The TOP network was coupled to the MAP network by a gateway. Thus even at such an early stage, an exemplary application in the shape of the technical execution of an order by data processing means from order procurement to manufacturing could be implemented.

Immediately after the Autofact fair, TOP 1.0 was defined. The American TOP User Group was founded in 1985. In Europe, SPAG (Standards Promotion and Application Group, consisting of 21 vendors) which was founded in 1983 as an organization of system vendors has influenced the development of TOP. On the user side, the European organization OSITOP was founded in 1987.

TOP 3.0 was introduced together with MAP 3.0 at the Enterprise Network Event 1988 in Baltimore and firmly established afterwards.

As already mentioned, TOP - like MAP - is based on the ISO/OSI reference model. From the very beginning of the development it was attempted as far as possible to achieve an identity of TOP and MAP protocols, so only the layers 1, (2) and 7 differ with regard to the applied ISO protocols, due to the different application requirements.

The currently valid version 3.0 is described below.

Layers 1 and 2

On layers 1 and 2, a coaxial cable as baseband with the CSMA/CD medium access method (ISO 8802/3) was chosen for TOP (ETHERNET). The data transfer rate is 10 Mbit/sec. Up to 1025 stations can be linked to the network.

The reason for this decision was on the one hand the maturity of the involved technology and on the other the high number of available ETHERNET products. More than 200 companies all over the world manufacture ETHERNET products. The stochastic access method CSMA/CD was chosen with reference to the wide range of applications in the office environment, where guaranteed transfer times are usually not as important as in the manufacturing area.

In TOP 3.0, a twisted two-wire cable as baseband (Token Ring ISO 8802/5) with a 4 Mbit/sec transfer rate or a coaxial cable as broadband with the CSMA/CD access method and 10 Mbit/sec can be alternatively applied. The LLC sublayer of layer 2 corresponds to the alternative in MAP 3.0: ISO 8802/2 class 1.

Layers 3 to 6

Layers 3 to 6 are nearly identical to MAP 3.0.

Layer 7

In TOP 3.0, apart from the association control service element (ACSE), file transfer and management (FTAM), directory services and network management, also the message handling system (corresponds to CCITT X.400) and virtual terminal (VT) are available, thus responding to the requirements of office applications.

Corresponding to the MAP development, the services available on this layer are to be enlarged step by step (e.g. by remote data base access).

Like MAP, TOP includes application interfaces (Application Program Interfaces).

As a very important point for applications in the office area, the connection to public networks, eg. via CCITT C.25 (layers 1 to 3) on layer 3 or CCITT X.400 (layers 6 and 7) on layer 7, is implemented. Other local networks can be connected via bridge, router, or gateway, depending on their type. Due to the corresponding specification from layer 2 onwards, TOP and MAP networks can be connected on layer 2 by a bridge.

From TOP 3.0 onwards, the compliance of TOP products with the given specifications can be tested by the test centres which were established for MAP.

The ESPRIT project CNMA also influences the further TOP development, as it has an effect on MAP.

(Further literature: Bauer 1987, CIM 1986, Kommtech 1986, Suppan-Borowka *et al.* 1986, Lukasik 1986, Pimentel 1990.)

Table 4.3 Specification of the ISO/OSI layers in TOP 3.0 (bold characters: identical with MAP 3.0)

Layer	TOP 3.0 specification
	Application Program Interfaces (API)
Layer 7: application	ISO MHS; CCITT X.400 (Message Handling System) VT (Virtual Terminal) DIS 9041 **ISO-FTAM DIS 8571** **ISO-ASCE DIS 8649/1-2** **DIS 8650/1-2** **Directory Services** **Network Management**
Layer 6: presentation	**ISO-Presentation DIS 8823**
Layer 5: session	**ISO-Session Kernel IS 8327 duplex**
Layer 4: transport	**ISO-Transport Class 4 IS 8073**
Layer 3: network	**ISO-Connectionless Internet DIS 8473**
Layer 2: data link	**ISO Logical Link Control DIS 8802/2 Typ 1** - ISO-CSMA/CD DIS 8802/3 (Coax) or ISO-Token Ring DIS 8802/5 Two-wire cable
Layer 1: physical	ISO-CSMA/CD DIS 8802/3 Baseband (10 Mbit/sec.) or Broadband (10 Mbit/sec.) oder ISO-Token Ring DIS 8802/5 Baseband (4 Mbit/sec.) - Two-wire cable

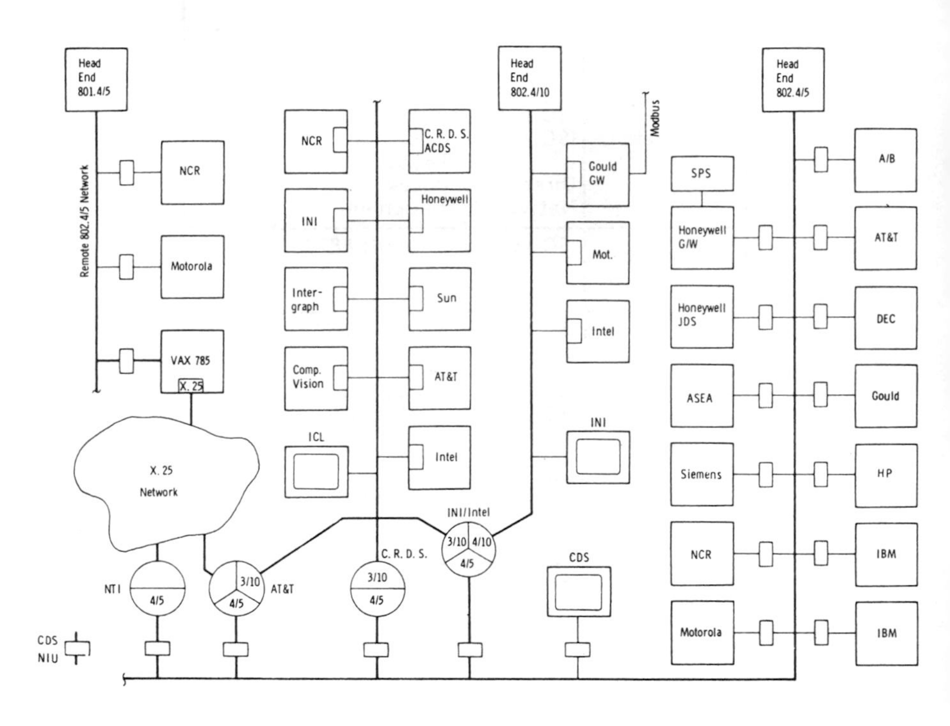

Fig. 4.13 MAP/TOP demonstration at Autofact '85 in Detroit (CIM MANAGEMENT 2 (1986) 3, p. 80).

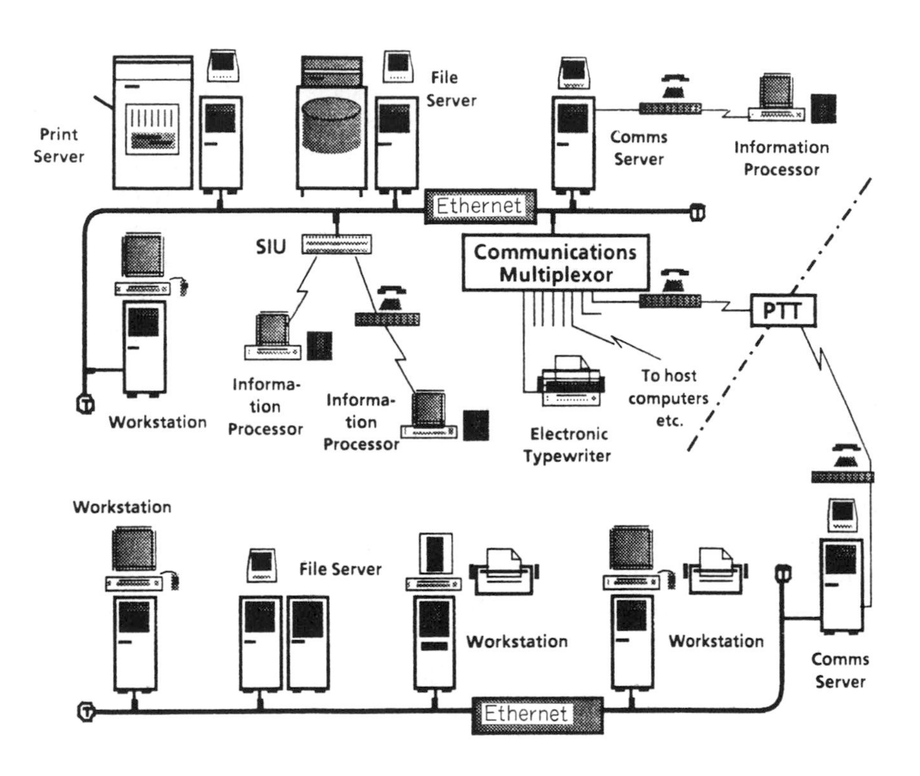

Fig. 4.14 Example of a network in the office environment.

4.4.3 Fieldbus

In the manufacturing area, the fieldbus forms a downward enlargement within the system hierarchy of the application area of MAP,. i.e. of EPA or Mini-MAP (cf. Fig. 4.12).

Standardization efforts aiming at an international standard have been under way for some time. An international standard which will not be defined earlier than in 1994 is currently being elaborated by IEC TC 65 SC 65C/WG6 (IEC - International Electrical Commission). This standard is influenced by national standard proposals or standardization projects such as PROFIBUS (Process Fieldbus, Germany, preliminary standard 1988, 1990), MIL 1553 B (Great Britain, standard), and ISA SP 50 (ISA - Instrument Society of America, USA, preliminary standard 1990), as well as the European EUREKA project (cf. for instance Borst *et al.* 1988) which mainly addresses the problem of intrinsic safety.

The fieldbus has the task of establishing open communication between elements of the so-called peripheral microelectronics at installations and machines (sensors, actuators, measuring transducers, drives, etc. and, on an intermediate level, also programmable logic controllers (PLC)). The demands placed on the fieldbus result from the types and the tasks of the communicating elements and the overall system. Some essential requirements are mentioned below.

- Data exchange between the elements at the installations or machines has to be conducted within milliseconds, i.e. in real-time. The volumes of data which are transferred are small, only several bytes.
- In comparison with computer installations, the communicating elements are low-cost products. Communication solutions have to show a realistic correlation with their price (connection costs of less than 100 DM per node).
- Elements such as actuators, etc., need auxiliary energy to operate which has to be supplied via the fieldbus.
- The length of the bus can be several kilometres.
- The bus must be able to process the different data rates of the elements.
- High fail-safety requirements must be met, especially in industrial process automation.
- The possibility of connecting a large number of different elements must be given (in extreme cases up to several thousands, though this implicitly contradicts real-time requirements).
- Connections to other networks must be possible (gateway, router, bridge).

To give an example, the Geman fieldbus project which was started in 1987 sponsored by the German Ministry of Research and Technology (and its follow-up project, started in 1990) in which 21 industrial enterprises and institutes are co-operating in the development and testing of a fieldbus standard, is described in the following paragraphs.

The fieldbus development within this joint project is based on the PROFIBUS concept (Process Fieldbus, DIN V 19245, part 1) which was jointly elaborated by Bosch, Klöckner-Moeller, and Siemens. DIN V 19245, part 1 includes the protocol definitions for layers 1 and 2 of the ISO/OSI reference model.

The concept itself is also based on the ISO/OSI reference model.

The physical medium is a twisted two-wire cable with a transfer rate of 500 Kbit/sec at 200 m (90 Kbit/sec at 1,200 m) (RS-485). The number of active or passive participants is limited to 32 (with a repeater a maximum of 122 can be attained). Commonly available standard integrated circuits are used.

On layer 2, a token passing protocol is defined.

Layers 3 to 6 of the ISO/OSI reference model are empty.

Layer 7 of the fieldbus will be defined in part 2 of DIN V 19245 in the near future. The concept of the protocol FMS (Fieldbus Message Specification) closely resembles the structure of MMS developed for MAP (cf. Katz *et al.* 1989).

The development steps and contents of the joint project closely resemble those of MAP and TOP:

- elaboration and implementation of fieldbus link control (FLC) and user protocols and interfaces;
- development of test environments and compliance tests, certification;
- construction of maintenance and diagnostic devices;
- demonstration at fairs (for instance the demonstration of the 'Unified Fieldbus' at the ISA-Show '89 (Philadelphia) and INTERKAMA '89 (Düsseldorf) at which mainly the vendors engaged in the EUREKA project participated);
- provision of information for vendors and users, reporting on the project's progress.

Within the framework of the project, it is intended to apply the fieldbus in the two application areas of manufacturing automation and industrial processing (cf. Rake *et al.* 1989), and here especially in building automation. A further application area mentioned by Pfeifer *et al.* is the communication of electronic systems in automobiles (Pfeifer *et al.* 1987). Further applications may include aeroplanes, ships, etc. In manufacturing automation, fieldbus systems may be used for (cf. Pfeifer *et al.* 1987):

- Statistical process supervision and control:
  During the whole manufacturing process, the relevant product data are gathered and evaluated by a diagnostic system with regard to product quality, machine and process performance, etc.
  The fieldbus serves as a serial shuttle bus. With the help of the bus, also set values for the manufacturing process can be transferred which were derived from the results of the evaluation process (remote control, event oriented interrupt processing).
- Combining several PLCs of one installation to a network:
  An assembly line is jointly controlled by a number of PLCs which can be tuned to each other by a central process control system. Error diagnosis and recovery are facilitated, and re-tooling, i.e. changes of the installation, can be performed more easily.

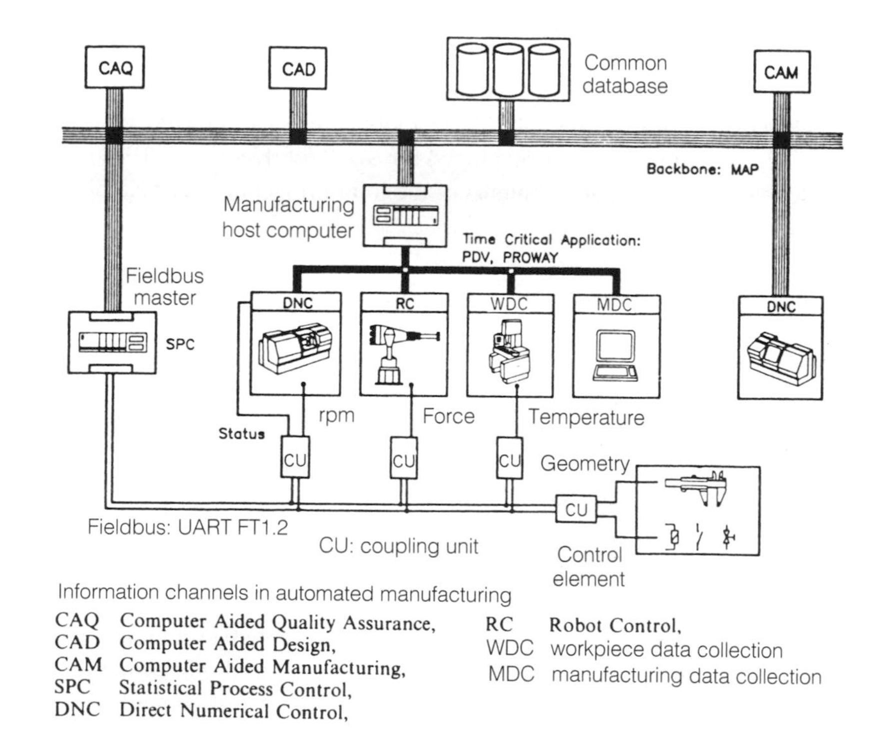

Fig. 4.15 Information channels in automated manufacturing (Pfeifer *et al.* 1987).

Primarily because of the ongoing digitization of measuring and control technology, the fieldbus with digital interface assumes an increasingly important role.

Due to the different requirements arising from different application areas, and also with regard to the resulting system overhead for some areas in case of a maximum consensus, the question whether there will be one all-embracing fieldbus standard or rather several specific standards cannot be answered yet.

Currently, only four fieldbus solutions are available on the market.

(Literature: Pfeifer *et al.* 1987, ATP 1988, Armitage *et al.* 1988, Wood 1988, Borst *et al.* 1989, Rake *et al.* 1989, Katz *et al.* 1989, Engel 1990, Henn 1990.)

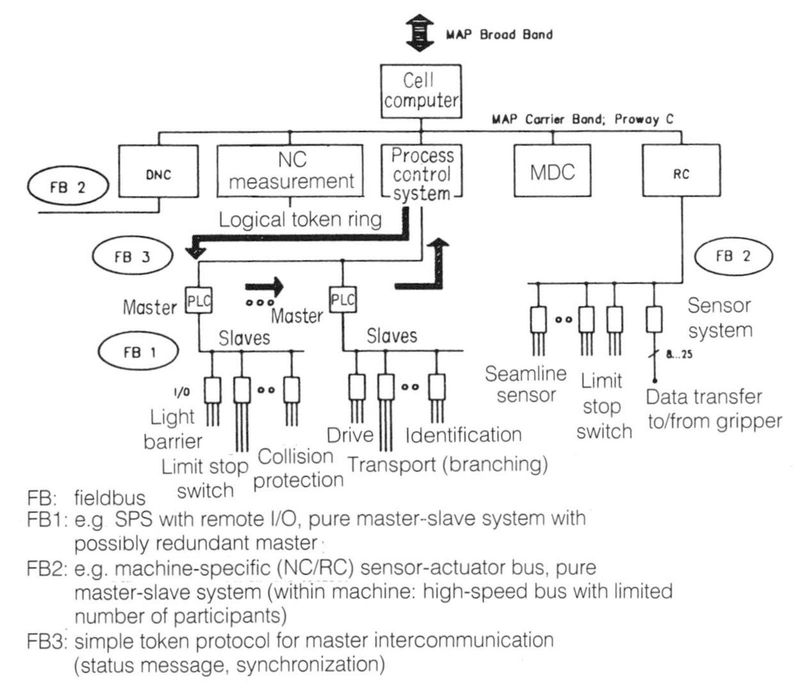

Fig. 4.16 Fieldbus applications in production technology (Pfeifer *et al.* 1987).

4.5 ISDN

ISDN (integrated services digital network) as a universal, service-integrating, digital telecommunication network will play a major role in the future development of CIM.

This is especially relevant for the integration of locally distributed applications and also for the increasingly important integration among different companies.

Until now, only few attempts at such an integration among different companies have been made, concerning the automobile industry and their suppliers.

The hitherto divided public communication networks, the telephone network and the integrated data network (IDN), together with their incorporated services (telephone, telefax, videotext, data communication, Datex-L, Datex-P, telex, teletex), will gradually be digitized and integrated into the unified network ISDN.

In the first place, ISDN provides the user with a unified basic connection to the network services. New services such as ISDN telecontrol, e.g. for process data transfer, ISDN textfax or the ISDN image transfer can be expected. The conventional services will become more comfortable, faster, and sometimes cheaper.

It is possible to differentiate between ISDN narrowband (S-ISDN) and ISDN broadband (B-ISDN). S-ISDN is currently being tested in two field studies by the German Bundespost and its nation-wide introduction is projected. It relies on the existing copper wires of the telephone network as the physical medium. The net tranfer rate of S-ISDN is 144 Kbit/sec (2 channels with 64 Kbit/sec each plus 16 Kbit/sec for the control channel), in contrast with the conventional transfer rate of 9600 bit/sec. The topology is a hierarchy consisting of stars. The connections are done via line exchange.

The application of B-ISDN is projected for the 1990s. In addition to the S-ISDN channels, two channels with a transfer rate of 2Mbit/sec (H1) and even 140 Mbit/sec (H4) will be available. This will, for instance, enable the transfer of moving images. The attainable data transfer rates correspond to the cycle times of the central processing units of universal mainframe computers currently available on the market. The physical medium will be glass-fibre cable. On behalf of the German Bundespost, B-ISDN is currently specified and implemented by the company DETECON in Berlin in the framework of the BERKOM project (Berlin Broadband Communication System). Pilot applications will also be implemented in the course of the project, focussing on the CAD-CAE-CAM area (cf. BERKOM 1987a, BERKOM 1987b).

Messer mentions several examples of the impact of B-ISDN on CIM (cf. Messer 1988):

- Due to the exceedingly high capacity of the channels, B-ISDN enables the use of simpler and more facile central databases without run-time losses.
- Locally distributed expert systems which may have a high dialogue intensity can communicate via B-ISDN from locally distributed functional areas within a company.
- Real-time controllers can be implemented more easily.

In co-operation with the BERKOM project, a private information technological infrastructure centring on B-ISDN will be implemented at the Technical University of Berlin (cf. TUBKOM 1987). Apart from the broadband communication infrastructure, applications of broadband communication will be developed in a second project. Here are some application projects with relevance to CIM which are to be addressed within the framework of TUBKOM:

- development of a communication system for integrated computer aided engineering in broadband ISDN;
- distributed design in mechanical engineering;
- fast data communication in the CAD and graphics area;
- distributed CIM structures;
- company internal integration of broadband ISDN services;
- company internal broadband integration of CAD data in documents.

A broad, industrial application of B-ISDN will not be possible before the mid-1990s (Felts 1988, Stallings 1989).

4.6 Summary

Local area networks are a means of enabling the communication between different data processing applications under the conditions as they are given in a real manufacturing company.

The technologies of local area networks differ mainly with regard to topolgy, transfer medium, transfer technology, and medium access method. The characteristics of a local area network are specified depending on the specific application area and its specific requirements.

In order to ensure open communication between very different types of computers or applications, ISO designed the OSI reference model in which the communication between open systems is represented by seven layers. The ISO/OSI reference model only gives a functional predetermination of the specific implementation of the layers.

For communication between partial networks, network connections are available. Depending on the compatibility of the protocols used on the single layers of the partial networks which are to be connected, repeaters, routers, bridges, and gateways can be applied.

The communications structure of manufacturing companies can be attributed to levels of a hierarchy (i.e. planning, control, and process level). Due to the different tasks of the single levels, different types of computers or intelligent systems are applied, and there are specific requirements for communication between these systems. These different perspectives formed the starting point of the standardization projects MAP and TOP which comply with the ISO/OSI reference model, and the fieldbus project.

MAP, which was initiated by GM, mainly covers the control level with the computers located in the manufacturing area. The MAP concept is based on a backbone network which links single automation cells with each other. These cells themselves can be networks. MAP (together with TOP) was publicly demonstrated for the first time at Autofact '85.

Because of its elaborate protocol structures, MAP itself cannot meet real-time requirements. Therefore reduced architectures (EPA, Mini-MAP) for sub-networks were defined on which unified or similar applications can communicate.

MAP version 3.0 was introduced in 1988. It is planned to remain valid for six years.

Problems for the user are mainly caused by the fact that until recently only few products complying with MAP were offered on the market. Since MAP 3.0 this situation has improved. Connection costs remain relatively high, and so do the installation efforts as well as the required qualifications on the part of the installation personnel.

A first implementation step for instance via Mini-MAP is beneficial.

TOP, which was initiated by Boeing, forms an enlargement of the application area of MAP at the upper level of the system hierarchy. The different applications in the design and administration area can be connected with each other. TOP version 3.0 has been valid since 1988.

The developments of MAP and TOP are being conducted in parallel. Both are based on the ISO/OSI reference model; their layers are filled with sometimes differing protocols as the requirements are not identical. It is attempted to achieve the highest possible level of identity.

For communication of and with sensors or actuators, it is planned for the lower level of the system hierarchy to enlarge MAP (and EPA, Mini-MAP) by the fieldbus. Development activities aiming at a standardized type suitable for open communication have only recently been started.

An international fieldbus standard will probably not be available before 1994.

Due to its very high transfer rates, ISDN, and especially its broadband version, will play a major role especially for the CIM area from the mid-1990s onwards as a universal, service integrating, digital telecommunication network. Possible applications are the integration of locally distributed structures within a company, but also the integration of areas among different companies.

Chapter 5

Examples of the coupling of CA applications

Viewed from the perspective of process organization, the coupling of CAD, CAPP and CAM systems tends to effect a vertical integration with regard to the information flow of product related data (cf. Fig. 1.6 and Fig. 5.1).

Through repeated electronic/data technological utilization of the product models elaborated by the design department, the work-load and the throughput times of those functions which precede the actual manufacturing process can be reduced. The quality of planning can be improved, for instance by avoiding errors and inconsistencies which occur during the repeated generation of the basic data by different systems. Finally, coupling leads to a simplification of the information flow.

Two main coupling methods of CA applications will be discussed in an exemplary manner in this chapter (for the coupling of CAPP and CAM see Section 3.3).

According to the definition given in Chapter 1, CAPP systems cover two major task areas: on the one hand routing generation, and on the other hand NC programming in a broader sense (NC, RC, PLC).

As for NC programming, geometrical data from the CAD area alone are not sufficient; data from the routings may also be necessary. First the coupling of CAD and computer aided routing generation and subsequently the coupling of CAD and NC programming will be discussed.

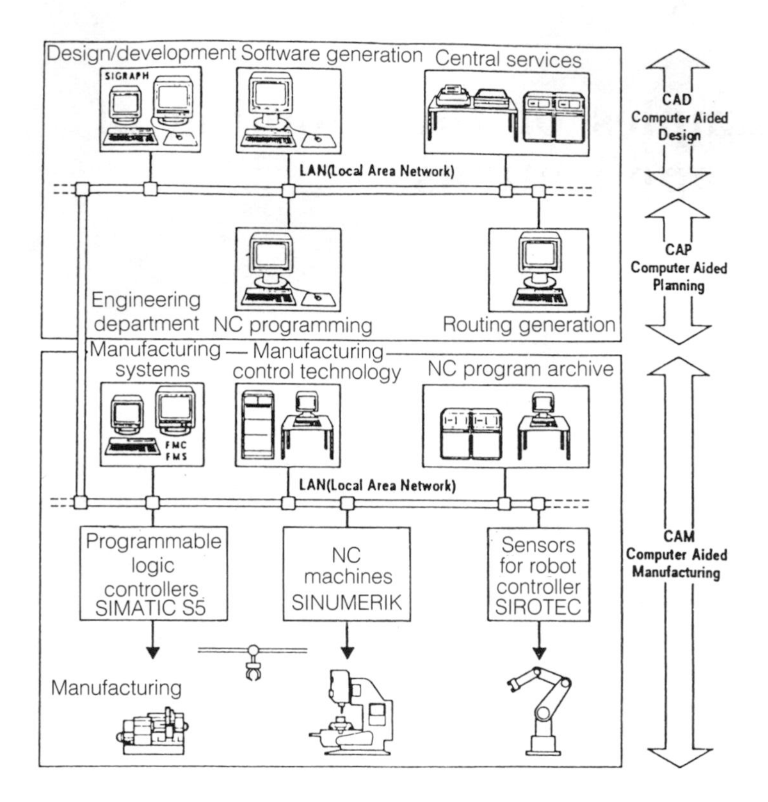

Fig. 5.1 Coupling of CA applications (courtesy of SIEMENS).

5.1 Coupling of CAD and computer aided routing generation

During the design process, essential characteristics with relevance to the manufacturing process are determined. Routing generation is an application of the geometry generated during the design process, mainly under constraints of production technology and economy. Input items for routing generation are technical drawings and manufacturing bills of material.

In the following a situation as it frequently occurs in current practice is outlined.

The design process in the tool construction area of a major enterprise is conducted using a standard CAD system running on a mainframe.

Data processing support for routing generation is achieved by individual software on personal computers. The routing generation module (RGM) is an interactive program. The input for the RGM is extracted by the process planner from copies of design drawings generated with CAD. With the help of these drawings the process planner decides which working operations the workpiece must undergo. Depending on the symmetry of the workpiece, certain technology modules are selected.

According to which technology module has been selected, the system calls for certain process-specific inputs.

All modules need the following information:

- wage group;
- workplace;
- department in charge;
- dimensions;
- set-up times;
- kind of material;
- surface quality;
- tools;
- special kinds of treatment;
- manufacturing process description; and
- quantity.

Some of these statements can be extracted from the design drawing. The others can be provided by the planner using his expertise, or have to be gathered from records or other data sources (material catalogues, resource files, auxiliary resource files, nomographs, timeframe tables and functions for the determination of lead times). A further possible means of support for the planning engineer which is not implemented in this example would be the search for similar parts with access to their routings already generated in the database of a PPC system, or the selection of a specific routing from a set of so-called standard routings which are valid for a certain family of parts (group technology). Via the technology modules, data for the following operation sequences are firmly determined:

- cutting to size;
- planing;
- turning;
- milling;
- coordinate drilling;
- coordinate grinding;
- engraving;
- eroding;

- finishing;
- grinding - round;
- grinding - geometrically;
- preparation; and
- heat treatment/blasting.

All other working operations have to be input via an empty template. The working operations which cannot be generated by technology modules include:

- bore milling,
- painting, and
- welding.

The processing of a routing is a six step procedure:

1. Input of head data.
2. Generation of the operation sequence data using the technology modules in the case of predetermined working operations, or empty templates in other cases.
3. Output of the routing.
4. If necessary, alteration of the routing with the help of alteration programs.
5. Storage of the routing on a personal computer.
6. Transfer of the routing to the mainframe.

At present, the process planner has to re-input the following data which were already input into the system by the design engineer:

- drawing number,
- designation,
- DIN format of the drawing,
- position number, and
- quantities.

Apart from these data which are explicitly stated, the process planner has to transfer a number of other statements from the drawing, such as in some cases the surface quality. In many cases, the relationship between the contents of the drawing and his input is clear-cut.

Often, though, no direct transfer of these statements is done. The process planner to a high degree draws on his knowledge, his experience and his spatial imaginative faculty for transferring the data which are scattered all over the drawing to the RGM.

By coupling CAD and computer aided routing generation, information necessary for routing generation which has been processed by the CAD system should be directly transferable to the CAPP system as its input data.

Actual practice within a company leads to numerous obstacles to an implementation of this concept.

One of them is constituted by the fact that the majority of CAD systems available on the market today are designed only for drawing or model generation. The model is stored using geometrical elements such as dots and lines. A logical connection leading for instance to the recognition of production technological elements cannot be made (Hüllenkremer *et al.* 1985).

Another aspect is that the programs supporting the generation of routings are more often than not individual solutions, as the routing generation to a very high degree is influenced by company specific circumstances and requirements. In many cases, both systems are running on entirely different hardware architecures. Therefore well established and tested instances of coupling CAD and computer aided routing generation are rarely available on the market.

Even though no satisfactory solutions have been developed yet, there are some approaches discussed in the literature.

The general possibilities of coupling can be seen in Fig. 5.2.

For technologically advanced coupling, the generation of production technological logic elements as part of the process of creating construction drawings is of prime importance. In this respect, today's systems are limited to special variant parts or generally to turned parts. In the following, two systems are briefly addressed.

Hüllenkremer *et al.* (1986) describe a system which automatically or by interaction at the CAD screen refines the geometry of the CAD system with coupling software and stores it in a special 'geometry definition file'. From this file, data are transferred to the routing generation system and again refined in the process of routing generation by a geometry planning logic program, i.e. for instance the shape of a turned part adopted from the CAD system is decomposed into production technological elements such as cylinder, recess, etc. Based on these data, routings can be generated with computer support. With this type of coupling, CAD and CAPP may be installed on different computers. A FORTRAN interface and, to a certain degree, also the EXAPT interface have proved to be suitable for this coupling.

A further system only applicable to parts with rotational symmetry is the PROPLAN expert system (for the following cf. Mouleeswaran *et al.* 1986).

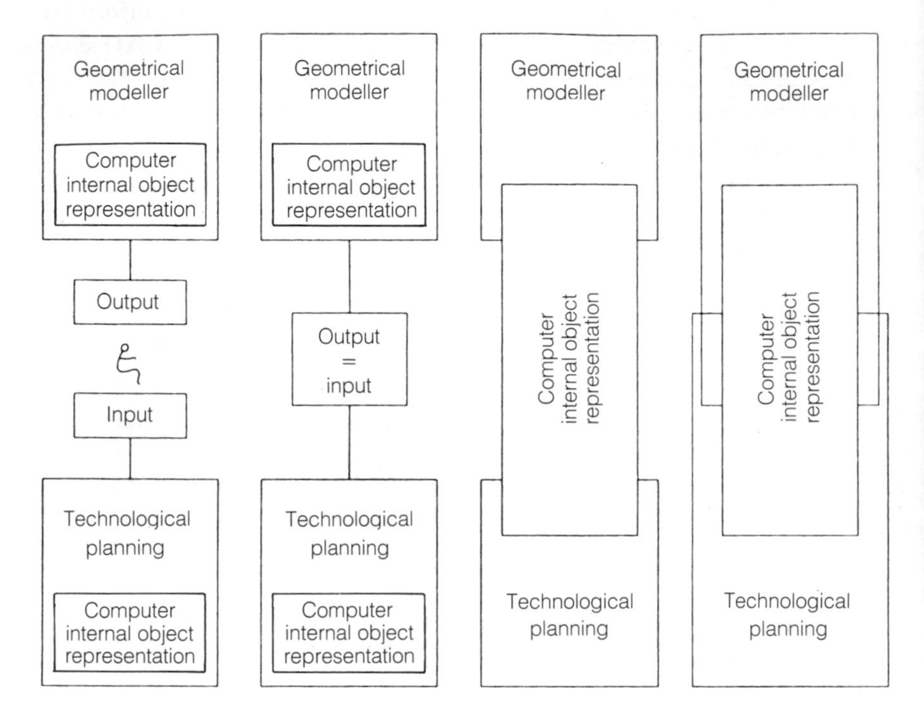

Fig. 5.2 Possible connections of geometric modeller and technology planning (Norsk Data 1986).

The name PROPLAN is an acronym of the term 'process planning'. In the USA, according to the International Federation of Information Processing (IFIP), process planning is a subtask of production planning and control. Thus process planning has the task of converting design data into process instructions.

Within the limits of their capabilities, machine tools have to be adjusted in such a way that a workpiece which is to be processed assumes after the operation sequence the intended final state of manufacturing as it was laid down in the design.

This problem is characterized by the existence of:

- a high level of complexity, and
- a large technological knowledge base

A precise knowledge of materials, machine operations, and external knowledge of the performance and the robustness of machines and materials are a precondition. Furthermore the group technological classifications of already manufactured parts must be taken into account. From the perspective of CNC technology, PROPLAN fulfils the tasks of:

- the path finding measuring system,
- diagnosis,
- error compensation,
- tool supervision and control, and
- automatic measuring and process control.

It operates in the following way.

Following the input of a rough design of the part which is to be manufactured, a wireframe model of the part is generated in connection with a CAD database. After final alterations of the rough wireframe, a solid model can be generated.

In an additional operation, the dimensions of the part geometries are defined, too. If the solid model corresponds to the intended part with rotational symmetry, the operation sequence for the part which is to be manufactured can be generated either automatically or interactively. Optionally, the single operations can even be graphically generated in detail with the use of graphic tools, whereby the workpiece in a simulation is manufactured on screen. Recommendations with regard to suitable machines and technological parameters are given. Furthermore the system explains its procedure for conducting the process planning. Here PROPLAN makes decisions on:

- the assignment of machines and operations, and
- which tools,
- which coolants,
- which cutting depth, and
- which cutting speed

are to be applied.

It must be mentioned that PROPLAN is no more than a prototype and until now has only been implemented for parts with rotational symmetry.

Judged by aspects of coupling capabilites, this system shows the highest state of development. CAD and CAPP are integrated into one application.

In current practice, the definition of production technology information during the design process which afterwards can be directly applied for routing generation requires that CAD and computer aided routing generation are performed in one and the same application. Thus routing generation may be implemented as a subroutine of the CAD system, or design and routing generation may run as parallel processes and parallel representation on screen is achieved with the help of window technology.

As a consequence, functional integration requires an enhancement of the qualifications of the design engineer to enable him to fulfil tasks formerly assigned to the process planner, or vice versa.

Standardized interfaces for the coupling of CAD and routing generation currently do not cover the relevant concerns of routing generation (cf. Section 3.2). In a future practical application of STEP, PDES, or PDDI, the possibility of transferring product definition data will support this interface to a much larger extent than now.

Real integration is only achievable with the development of an integrated product model which contains the data generated by the CAD system and further data produced by routing generation. Alterations of the routing caused for instance by the application of new utilities which may have an impact on the design can then be put at the disposal of the CAD system without difficulties.

The system TechMo which is at present being implemented as a prototype is based on this approach of an integrated product model consisting of a technical, a geometrical, and a technological partial model which are represented in a common non-standard database (Hübel *et al.* 1990).

5.2 Coupling of CAD and NC programming

The coupling of CAD with NC programming is already relatively far advanced. In a survey conducted as early as 1986 among CAD vendors (Latz 1986), all 16 systems examined enabled the transfer of product definition data to NC programming by generating suitable part programs in an NC programming language.

The utilization of geometrical data provided by the CAD system is the main concern of NC programming. However, technological data such as tolerances, surface quality, or materials also play an important role.

Design and NC programming are conducted separately by different departments of a company. The applied hardware and software systems run independent of each other with different data and storage structures and on different computer systems. Until recently, NC programming was performed with the help of design drawings, even when these had been generated with a CAD system.

The direct utilization of CAD data for NC programming effects a considerable improvement of efficiency, due to the significant congruency of common data volumes.

Programming is facilitated because the labour intensive re-input of data can be avoided. The probability of errors can be reduced, as the danger of transfer errors as a result of renewed data input is lessened.

The possibility of connecting CAD and NC programming is practical only if NC programming is conducted automatically. This contrasts with manual NC programming, which is performed by coding the control programs in machine language (according to DIN 66 025) on an input device such as for instance a tape perforator, or by a machine engineer who in a guided dialogue at a terminal directly inputs the control program into the NC machine (workshop programming).

Automated programming is mainly done in higher programming languages such as EXAPT, APT, or their dialects. The programs are interactively input into the computer by an NC programmer using an editor. The advantage lies in the computer aided extraction of data about machines, tools and materials, and the automatic generation of tool motion paths, etc.

The coupling of CAD and NC programming itself can be achieved in several forms (for the following paragraphs cf. Milberg *et al.* 1987, Hellwig *et al.* 1983a, Hellwig *et al.* 1983b, Hellwig *et al.* 1985, Magill *et al.* 1988).

If NC programming is integrated into the CAD system, geometry transformation via an intermediate file into an NC oriented geometrical representation is unnecessary. Design and NC programming are performed on the same database and the same computer system. The NC programmer, using the geometry data, generates the part program in a graphically interactive dialogue at the CAD workshop. Thus he has at his disposal the same commands and graphic operational possiblities as the design engineer. The main task of the NC programmer is to describe the motion paths of the tools. These are linked to the geometry of the part by pointers; therefore, an alteration of the design effects an automatic change of the motion paths. During the generation of the tool motion paths, the tool parameters of the tools chosen by the NC programmer from tool catalogues stored in the CAD/NC programming system are automatically taken into account. Some additional technological information has to be provided by the NC programmer.

The integration is especially advantageous if the NC programming system has to achieve a high geometrical performance level, i.e. if geometrically complex workpieces, for instance with free-form surfaces, are to be processed.

Practical obstacles to integration arise from technical as well as organizational difficulties. Workpieces which cannot be designed with the CAD system cannot be

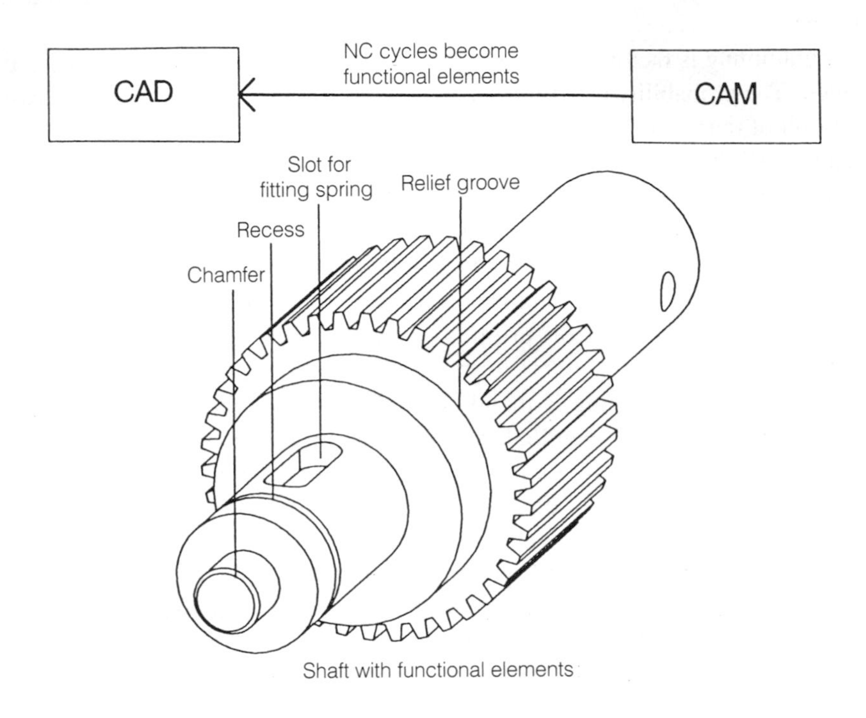

Fig. 5.3 NC cycles become functional elements (courtesy of SIEMENS).

programmed either. At present, the use of CAD stations as NC programming tools still remains a quite expensive solution.

Design and NC programming are assigned to one person. This leads to so-called job enrichment which in itself is quite beneficial. The lack of manufacturing know-how in the design departments, however, still forms a serious impediment to the implementation of this concept. A large number of technical problems, for instance the choice of suitable processing methods, problems of clamping and tension, correct selection of operation sequences and tools have to be solved. A further but perhaps less relevant solution would be to train NC programmers in design engineering.

In Fig. 5.4, the alternatives 1 and 2 are represented as integrated solutions. The difference between the alternatives is that in alternative 1 the information for a particular NC controller can be directly generated, while in alternative 2 a neutral intermediate format (CLDATA, cf. Section 3.3) is created, which with the help of postprocessors can be adapted to the various existing machine controllers. At present, because of its flexibility, alternative 2 is preferable, as DIN 66 025 only provides a basis for a unified structure of control programs, while the programs for the various controllers differ in details.

The alternatives 3 to 8 show CAD and NC programming systems as different program systems which are coupled with each other.

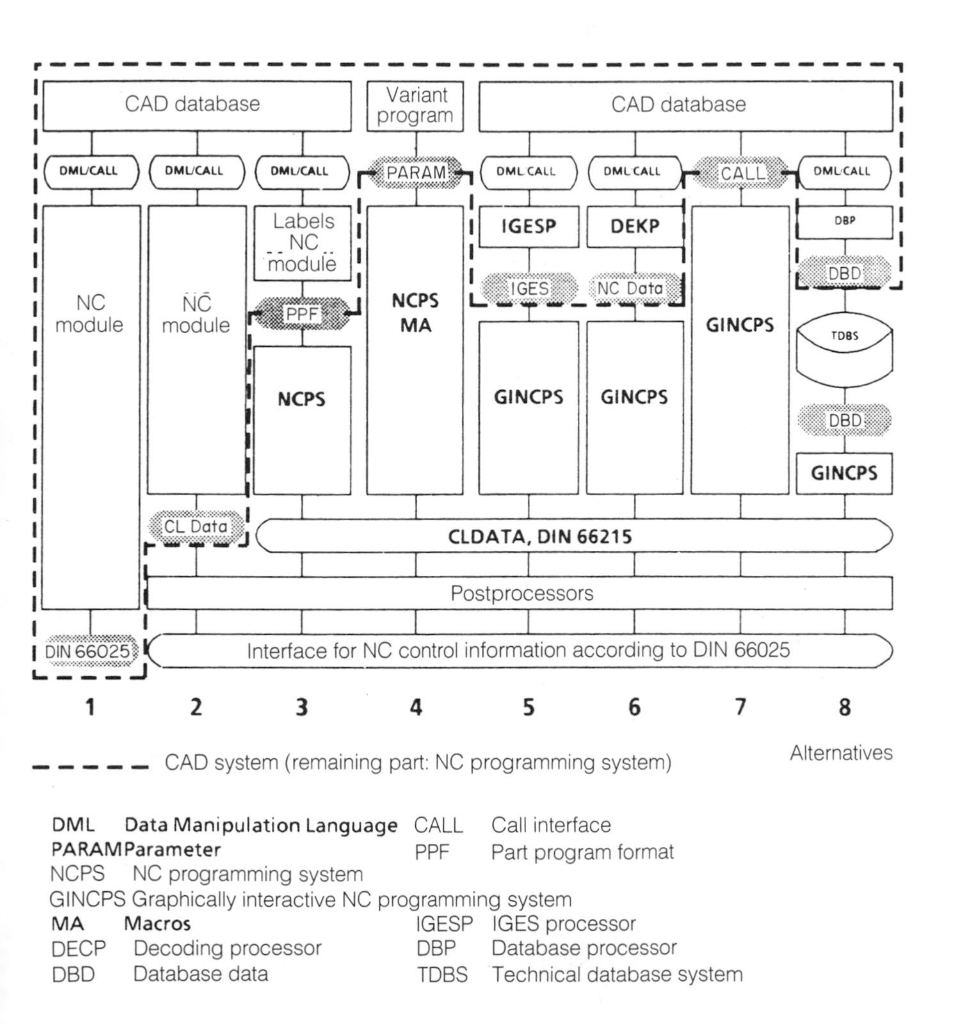

Fig. 5.4 Alternative possibilities of connecting CAD and NC programming (Milberg *et al.* 1987).

The required coupling of both systems is normally achieved by intermediate storage of the data generated by CAD and their processing in a converted form by the NC programming system.

Older CAD systems as they are used in practice often lack some of the preconditions necessary for coupling, i.e. a structured data transfer to an NC programming system, because of their basic structure. Older NC systems are non-interactive or do not always operate with a universally applicable NC language.

There are two general possibilities of coupling CAD systems and NC programming systems. They depend on the type of the applied NC programming systems and are characterized by the data structures of the coupling file:

- Language interface:
  The CAD system generates a file in which the geometry is stored in the desired NC language type (APT, COMPACT II, etc.) (language based NC programming systems: alternative 3).

- Data interface (descriptive interface):
  The CAD system provides the means to store the geometry data, e.g. with the help of an IGES postprocessor in a standardized file with IGES format (cf. Section 3.2.1) (graphically interactive NC programming system: alternative 5 - exemplary system: CADCPL) or as an individual solution in a non-standardized file (alternative 6).

This intermediate file is then read and processed by the NC programming system. If required, the input data are transformed into the desired NC language form and supplemented by other information, such as technological data and characteristics of materials, tools and machines. The result is usually a complete part program.

Pham summarizes the preconditions for coupling of CAD and NC programming as follows:

- automated NC programming, preferably with a universal language such as APT and graphical interaction;
- CAD and NC programming system have to use the same interface;
- data exchange has to be possible via tape, diskettes, hard disks, or optimally by on-line computer links (Pham 1982).

In alternative 3, the supplementation of the data by technological information can be performed by the NC programmer alphanumerically in interaction with the system by using an editor which is a part of the NC programming system. In contrast to fully automated programming, programmers do not have to generate the geometry definitions themselves.

Part programming with the help of a serially interactive system is more comfortable. The NC programmer has to add the missing instructions in a predefined sequence. For example, before programming the tool motion paths, the geometry has to be selected and transformed into a contour. The main advantage of this procedure is that the generated part program cannot include any syntactical faults. The semantics, however, can only be checked during a run on the NC programming system. Because of the defined serial input, the possibilities for later supplements or alterations of the part program are limited. Such alterations are easier to perform if the part program is generated interactively, allowing for corrections. Existing instructions can be changed without difficulty, and new instructions can be added.

A further variety (alternative 4) can also be covered by the term coupling, although it already shows some characteristics of integration, as, in consideration of the subsequent NC programming, it leads to changes of the design process which make it different from the normal design procedure conducted by the CAD designer. The designer has to assemble the model partially or completely by so-called shape elements. Apart from the geometrical descriptions, shape elements also contain manufacturing information about the part. The shape elements are put at the disposal of the design engineer as predefined macros. Because of this procedure, the data transferred to the NC programming system do not only contain geometry data, but production technology information as well.

The manufacturing information already stored for such standardized elements does not have to be generated once again by the NC programmer creating the part programs. Therefore the chance of errors is further reduced.

Graphical editors for part programming (alternatives 5 to 8) considerably simplify the work of the NC programmer. The most comfortable type of such an editor is the CAD system itself.

To give an example for a coupling of CAD with an NC programming system, in the following paragraphs the system CADCPL which can be attributed to alternative 5 shown in Fig. 5.4 is presented.

CADCPL receives the model data from different types of CAD systems and uses them primarily for generating NC programs with EXAPT. The system has a two-level architecture.

On level 1, the model data from the different CAD systems are tranferred into a standardized file (IGES) to enable their further processing independent of the CAD system used.

The CAD data are adapted by the system for further processing, if necessary in a dialogue with the user. This includes for instance the deletion of superfluous data, the selection of relevant data from the single geometry sets, the recognition of logical interrelationships between the single geometrical elements, the recognition of hidden elements, and the generation of symbolic names for individual geometrical elements (CAD adaptation module).

On level 2, part programming for the corresponding manufacturing method is performed. The single processing steps depend on the performance of the previously used CAD system and the NC programming task to be accomplished. The single steps can be for instance

- the assignment of tolerances, surface designators, etc., if this could not be done with the CAD system;
- the deletion of geometries irrelevant to the respective manufacturing method,
- the addition of new geometries relevant to the respective manufacturing method; and
- the coupling of geometries for the respective manufacturing method.

This is performed with graphical interaction using functions for geometry processing similar to those available in CAD systems.

Moreover, there are functions which enable users to view the EXAPT instructions they have generated in the preceding processing steps. With the help of editing functions they can add, change, or delete instructions.

The technology planning for all processing methods is performed with graphical interaction, too. After the completion of all processing steps, CADCPL delivers a complete, formally error-free EXAPT part program.

The essential disadvantage of this type of solution is that the IGES definition does not contain some elements of prime importance for NC programming. Also other IGES-specific difficulties (see Section 3.2.1) do occur and have to be taken into account.

Alternative 6 presents an individual solution for the interface file which is of sppecific relevance to NC programming. Its main disadvantage is the high development effort which has to be made for each pre-positioned CAD system.

Alternative 7 shows an arrangement where the NC programming system has direct access to the CAD database. Here the closed system architecture of the available CAD systems as well as the insufficient capability of current NC programming systems to access external databases of different types are difficulties still to be overcome.

Alternative 8 describes a further hitherto theoretical solution where both systems have access to a neutral external database. Non-standard database systems for this purpose are still in the research and development phase (cf. Section 2.4.2).

Coupling methods applied in practice are represented by the alternatives 3, 5, and 6.

The advantages which can be gained by coupling CAD and NC programming systems certainly depend on the development stage of the applied systems and also on individual preconditions of the given company environment. To give an example, some of the benefits which could be achieved in the Trumpf company are mentioned hereunder (cf. Pham 1982):

- reduction of the programming effort for sheet metal parts of 15 - 40%, depending on the complexity of the part and the processing method; for wire eroding it was as high as 70%, due to the low number of technical instructions required;
- avoidance of input errors as no renewed input of the geometry was required; and
- improvement of the information feedback between the design and NC departments.

At present, standardized interfaces do not sufficiently meet the requirements of routing generation and NC programming. Comprehensive product models which cover the logical connection of the CAD model and the NC model have yet to be developed. Primarily data of specific relevance to process planning such as design technological elements (e.g. borehole or slot) and their relations (e.g. boring images), shape and position tolerances with regard to type and dimension, dimension tolerances in the form of reference planes and points, or component joint data (for assembly) have to be transferable by the interfaces (DIN 1987c).

Interface standards which offer an enhanced support for the coupling of CAD and CAPP such as PDDI, PDES, and STEP, still exist either only as pilot implementations or are in the development phase (cf. Section 3.2, and in more detail with regard to the CAD-NC process chain, DIN 1989a).

5.3 Summary

Seen from the perspective of process organization, the coupling of CA applications is a vertical integration and concerns the information flow of product related data. Presently it is primarily a uni-directional coupling from CAD via CAPP to CAM.

The areas 'CAD - computer aided routing generation' and 'CAD - NC programming' have been discussed. The coupling of CAPP with CAM, i.e. the exchange of production technology data in automated NC programming, has already been standardized to a high degree and therefore was addressed earlier in Section 3.3. Still in the early phase of development is the coupling with robot controllers, and even more so with flexible assembly control.

In the partial area 'CAD - computer aided routing generation' only few approaches for coupling have yet emerged. A serious obstacle is the fact that programs for routing generation are mostly individual solutions due to the large number of company-specific requirements especially in this field. For coupling, a coupling processor using decision table technology in connection with a FORTRAN or EXAPT interface file has proved to be suitable.

Further possibilities are opened by expert system technology. The prototypical system PROPLAN is an integrated solution, as CAD and computer aided routing generation are combined in one application.

Currently, standardized interfaces for the coupling of CAD and routing generation do not yet meet the requirements of process planning.

Within the partial area of coupling CAD and NC programming systems, some adequate solutions have been implemented. There are three basic methods of coupling:

- CAD systems with an integrated NC programming module;
- CAD systems which provide data in a standardized format (e.g. IGES); and
- systems providing data in formats adapted to special NC systems (EXAPT, APT, etc. - language interfaces).

Standardized interfaces should be better adapted to the requirements of NC programming. Nevertheless, they can already be applied in their present state.

Truly integrated solutions will only be feasible when all CA applications operate with a common logical product model.

Chapter 6

Coupling of PPC systems and CA applications - example: PPC - CAD

The most advanced data processing systems in the manufacturing area are PPC and CAD systems, as typical applications. Attempts to integrate both systems, and also to integrate PPC and other CA systems are still rare; if they exist, they mostly concern the PPC system (with its partial function 'routing management') and the CAPP system (with its partial function 'routing generation').

PPC vendors on the whole have only recently begun to address the issue of geometry/technology oriented applications. Although users enforced co-operation with competent vendors and developers in the CAD/CAM area, the purchased systems were hardly integrated into their already existing systems with regard to the database and often even the hardware. In the CAD/CAM area, specialized software vendors cooperate with hardware specialists. The results of this co-operation are isolated CA systems. Requirements of the PPC area are hardly ever met.

At present, standardization approaches for a unified interface between CA applications and the PPC system are still only beginning. Standards for the transfer of data between CAD and NC programming are still in the development phase. The standardization effort directed at STEP can be viewed with some expectancy, although here also the coupling of CA applications and PPC systems is not the dominant issue. Now as before, the coupling of technical and administrative data processing is mainly achieved by specific solutions, and all the more so when in the management and administration area outdated software is used.

A coupling of PPC systems and CA applications should be bi-directional as the following examples clearly show.

A CA system not only accelerates and economizes the design process, it also stores essential basic data which are already being generated during the design (in the first place elements of the design bill of material); they can be transferred automatically to the PPC system and do not have to be laborously input once again.

Also in the reverse direction, a coupling can be beneficial: according to studies conducted by the VDMA (Verband Deutscher Maschinen- und Anlagenbauer - Association of German Machine and Plant Manufacturers), the design process has an influence on 70% of the costs of a product. It can therefore be regarded as meaningful or even imperative to supply the designer with information concerning the costs of the applied materials and also the manufacturing methods and costs emanating from the design. Topical information from the PPC area such as delivery dates, critical parts, purchase costs, and manufacturing costs, etc,. must be made available to the designer. An important issue in this context is 'cost calculation accompanying the design' (cf. Scheer 1990).

The coupling of a PPC and a CAD system will be described in the following; it serves as an example for the coupling of PPC and CA applications in general (cf. for the following Scholz *et al.* 1988).

6.1 Example of the coupling of PPC and CAD

Numerous data flows do exist between the CAD system and the PPC system in both directions. An intensive data exchange between both systems leads to an increase in data integrity within both areas and can have a beneficial influence on the cost structure of the products (Harhalakis *et al.* 1987, Meijer *et al.* 1989).

The basic CAD data come in the shape of construction drawings or the information which can be extracted from them. These are also data of the design bill of material. The drawings are stored in the graphics computer as graphic data structures.

Nearly all the functions connected with business adminsistration, from sales to shipment, use basic CAD data extracted from the construction drawings.

The PPC system supports the planning and execution of the manufacturing process.

Basic PPC data include bills of material, routings, and resources. This on the whole coincides with the common data structure of PPC systems: part master files and part structure files as information for the bills of material, operation master files and operation sequence files as information for the routings, and resource master files and resource structure files as resource data.

The bill of material contains information about the structure and the quantities of raw materials, components, and assemblies of which the products consist. The archetypical form of a bill of material is the design bill of material which is compiled by the design department. The designer is the first person to know whether the bill of material is complete and whether all parts of the construction have been designated correctly.

In the design bill of material, all parts of the whole design are gathered in a complete and systematic table. Each entry in the bill of material states the corresponding part number, description, sizes, drawing number, and required quantities. Basically, the bill of material is a verbal listing of the composition drawings.

Based on the design bill of material, all the other bills of material are generated: a single level part list for kitting, a summarized component list for purchasing, a complete cost build-up list of a major assembly, and a manufacturing bill of material (level by level explosion) as the basis for the PPC system. The essential data of the original bill are enlarged or transformed with further information concerning availability, purchase, or process planning.

Design bills of material are often arranged according to functions. Subassemblies and parts which are included in one generative structure are combined into a group. The functional arrangement is further solidified by the definition of functional macros necessary for the CAD systems. The administrative area, on the other hand, requires not a functional but a manufacturing oriented arrangement (Steinmetz 1991). Therefore the arrangement of the bill of material may have to be altered several times by the PPC system. In principle, the bill of material should always document the generative tree of a product or an assembly.

Analytical bills of material decompose the generative structure of a product starting from the final product into assemblies and single parts. Synthetical bills of material, starting from single parts or components, show the use of the parts. Synthetical bills of material therefore are also referred to as used lists and constitute an inverse bill of material which determines the quantities of each part to be incorporated into the superior assembly group. In the case of automated bill of material management, and if a suitable form of storage is given, the used list is, as it were, a by-product of the bill of material management. The used list generated by the PPC system forms an important information resource for the design engineer in case of design alterations.

The design bill of material is also one of the information resources used for process planning. The routing supplements the bill of material and determines which operations have to be performed in which sequence with which utilities and at which point in time for manufacturing the parts contained in the bill of material.

Bills of material also provide basic information for all functions related to production planning and control. The design bill of material as the first version of all bills of material within the system supplies essential information generated by the design area, i.e. the CAD, to the PPC system for the first time. In the shape of a verbal list of the composition drawings, it transports the basic contents of the construction drawings. Therefore the bill of material assumes a critical role in the data flow from design (CAD) to the administrative area (PPC).

From here on the generation of the bill of material is based on construction drawings stored in the graphic data structures of the CAD system. The essential information is located in the text fields of the drawing's and the component drawings' head sections. For several CAD systems additional software for finding and evaluating these head sections in the data structure file is available. Design bills of material generated by these means contain information about the product structure only to the extent to which it has already verbally been included in the head sections of the composition drawings.

Bills of material are of course not the only link between the geometrical/ technological and the administrative/operational data processing areas. Just to mention one example, process planning (CAPP) generates information about routings which the PPC area needs for capacity planning. Moreover, there are resource data from the CAM area.

Conversely, CAD itself requires information from the PPC system. PPC information about similar parts and drawings can curb the accumulation of special designs generated by the design area. Characteristics sections according to DIN 4000, as they are integrated into many of the more advanced PPC systems, are a means of supporting the search for similarities. This improvement of information resource management can ideally lead to a form of variant design. Information about standard parts can facilitate the use of standardized and repeat parts. In case of design alterations, used lists indicate which parts in which context are affected by the alteration.

By using PPC routing data, standard manufacturing methods can be taken into account during the planning and design process. Access to specifications concerning resources and workshops as well as available capacities facilitates the design process by giving the design engineer a view on the actual shop floor situation.

Cost calculations with regard to in-house manufactured or purchased parts and information about stocks and delivery dates supplied by the PPC system can assist the designer in making essential decisions during development and design which have an influence on the product's cost.

In development and design, procedural and descriptive interfaces play a major role (DIN 1987c).

Methods and results of the different stages of development and design are based on widely differing types of information. During the development of a product, product descriptive information is generated and refined, starting with a high degree of abstraction and becoming gradually more concrete. Consequently, highly dissimilar types of data have to be managed and exchanged within computer aided design itself and between CAD and other areas such as CAPP and CAM or PPC. They include product definition data such as

- geometrical/topological data (shape and dimensions of components),
- graphic data (graphic representations of components),
- technological data (physical and structural characteristics of objects),
- structural/associative data (for combining data into groups),
- functional data (functions and active principles),
- organizational data (systematics, aspects of the throughput of orders),
- aspects of communication technology,
- parameterizable data (variants), and
- procedural/methodological data (methods and rules for variants and in general).

Interfaces as they are currently known such as for instance GKS, IGES, or VDAFS are mainly confined to the exchange of geometrical and graphic data between two CAD systems. They have a limited capacity for transferring technological, analytical, and structural/associative data, and their applicability to all other data types is insignificant (DIN 1987c).

Apart from these product definition data, vertical and horizontal integration within a company requires the exchange of further data with development and design. These include production technology data (manufacturing, assembly, quality assurance), order related data, and logistic data. For production technology data there are the interfaces of more high-level NC/IR programming languages, CLDATA and IRDATA, which to some extent can also be misused for the transfer of geometrical data, although this should be avoided. The standardized interfaces, however, are only applicable for a uni-directional transfer between the CAPP and the CAM area; thereby the relationship between CAD data and product technology data is lost (cf. Chapter 3).

No interfaces are available which cover data about tool parameters and applicable manufacturing methods and machines, even though these data are of relevance for the design. For the transfer of order related and logistic data only user specific solutions exist (for more details cf. DIN 1989b).

Interface solutions which address the coupling of CAD and PPC are not yet mature. Concerning the coupling or integration of PPC with CAD systems, Hellwig and Hellwig (Hellwig *et al.* 1987) do not believe that universally valid solutions will emerge. If existing systems are used - and that seems to be their supposition - standardized solutions are usually inapplicable.

In future, Hellwig and Hellwig intend to include technical drawings in the PPC data set also, as they, together with the various manufacturing, assembly, and inspection operations, are being transferred to the different workshops of a factory. 'Drawings are operative resources in a wider sense, just as are NC programs and routings, which are necessary for order release. Therefore these resources must be kept in and managed by the PPC. The same applies to standards and instructions.' The drawings do not have to be alterable, it is sufficient to include them in a fixed form, a 'container' as it were, in the relational PPC database.

Usually, drawing management is performed on the databases of the CAD systems. The concept of CIM, however, calls for a unified and consistent data stock which may be physically decentralized, but nevertheless must permit the central management by one system.

Hellwig and Hellwig propose two alternative solutions for an interface between CAD and the drawing management system (DMS): if the DMS is running on a CAD computer, for the PPC area an access method via a search algorithm offering output has to be implemented, but without the privilege to make any alterations. If the DMS is located on the PPC computer, the drawings generated on the CAD computer must be reformatted (deletion of structures) and transferred to the PPC computer. There the reformatted drawings are stored and managed by the PPC database. For alterations, etc., these 'container' drawings are transferred back to the CAD system and restructured.

For the CIM commission (KCIM) within DIN (DIN 1987c), two interfaces for the coupling of CAD and PPC are of special relevance. The execution of design processes is included in the PPC order management with the help of an order interface. The bill of material interface constitutes one of the most difficult and important interface problems. Its characteristics are

- a large data volume,
- high data quality requirements,
- the requirement of topicality of data even in case of frequent alterations, and
- its frequent use for production planning and control.

The problem is further aggravated by the different structure of bills of material used in design and PPC, and also by the fact that only a certain part of the data they contain is jointly used by both areas. Often problems of competence arise between the departments involved which in certain cases prevent the use of a common database. According to DIN, the standardization effort must focus on the data and data records themselves rather than on procedures for data exchange between the systems.

For the CAD area, the CIM commission regards the following standardization issues as critical:

- transfer of standard parts: standardization and introduction to industrial applications;
- geometry interface: introduction to industrial applications;
- integrated product model: development first for the mechanical engineering area (STEP), and later also for electrical and electronic applications (EDIF);
- transfer of organizational data; and
- program interface: development and standardization.

Because during drawing generation an informational determination of the logical structure of the bill of material is effected, an integrated solution will make information extracted from the geometrical data available to the PPC basic data management. Conversely, the design will have access to information about bills of material contained in the PPC database.

Until now, a coupling of bills of material - if implemented at all - exists only in the form of individual solutions for specific user purposes or as non-standard solutions offered by certain system vendors. Nevertheless, these solutions are state of the art (cf. Hackstein *et al.* 1990 about the problems of individual solutions).

Many of the couplings currently applied are based on simple methods of data transfer. On the whole, these are special solutions which could be implemented without much effort. Often the software systems which are to be coupled use common memory areas of a mainframe which are employed by both systems for data transfer and to provide each other with access to information. Thus the data which are to be transferred must share the same contents and structures to a considerable extent in order to minimize the conversion effort.

If the data structures of the systems differ greatly, substantial adaptations must be performed. In this case it is helpful to locate the conversion algorithms in special interface modules which as components intersecting with both areas take over the intermediate conversion step.

Examples of coupling processors offered by system vendors are, among others, CADMIP (Computer Aided Design to Manufacturing Interface Program) by IBM which links the CAD system CADAM sold by IBM to the PPC system COPICS, and CADIS-PPS which couples the CAD system CADIS and the PPC system IS (both by Siemens). IBM and Siemens limit their approaches to supporting their own systems, as considering all the existing CAD systems with their many different features is hardly practicable.

For the transfer of information from CADAM to COPIS, the head section of a drawing plays a prominent role. The transfer is acccomplished by applying attribute technology.

In composition drawings, the bills of material of the single parts of the drawing are included; thus a complete bill of material can be compiled. There is still the possibility of manually adding more lines to this bill of material.

If the composition drawing contains several identical parts (e.g. screws), the complete number of these parts listed in the bill of material can be stated by adding a comma and a number after the part number.

Being an overlapping module, CADMIP features design functions (access to PPC basic data, preparation of bill of material transfer) as well as communication functions (data transfer) and production planning functions (reception and processing of the bill of material, retransmission of information).

The single functions are as follows.

Design function

- the preparation of production planning data,
- if available, the examination and use of PPC data in CADMIP in order to update the drawing,
- the provision of data for clearance,
- the clearance of engineering (design) data,
- the indication of transfer status,
- the transmission and reception of messages, and
- auxiliary function.

Communication functions

- the acceptance of designs from all designers,
- the processing of clearance requirements of the design,
- the admission of later release for manufacturing,
- the generation of interface files and their storage for access by production planning,
- the passing-on to design of replies and reactions by production planning.

Production planning functions

- the reception of cleared design data,

- the insertion into the COPICS product definition file of:
  - points, items, pieces, notes,
  - product structures, and
  - alterations of designs,
- the transmission of information back to design (acknowledgement of reception), and
- the indication of 'delta reports' in case of differences.

Fig. 6.1 indicates the coupling scheme of the CAD system PROREN of the company ISYKON Software Ltd. to the PPC system.

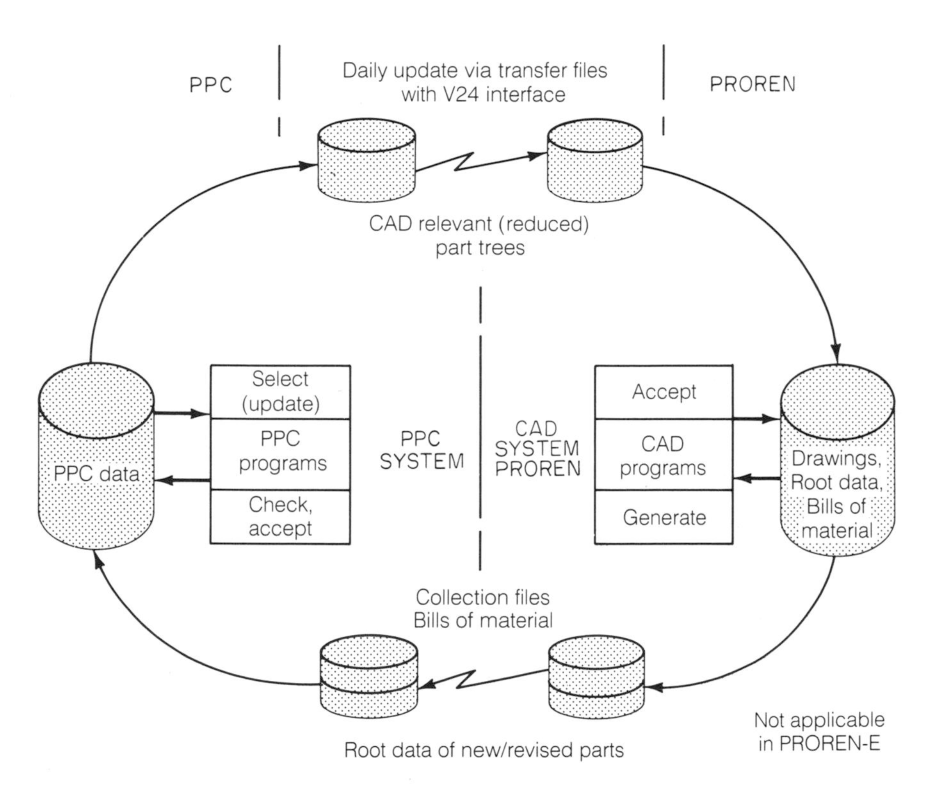

Fig. 6.1. Coupling scheme PROREN to PPC (courtesy of ISYKON).

On account of the crucial role the bill of material plays for the manufacturing areas, Eigner *et al.* (1986) examined integration via bills of material. As there are no standardized interfaces between CAD and PPC, the bill of material becomes the key item to be standardized between CAD and PPC, much the same as in the case of the IGES interfaces between CAD and NC programming.

A program package for the management of design bills of material on the one hand serves as a management and information system for the design engineer; on the other hand it is an important module for data transfer between both systems. Its flexibility allows the coupling of any type of CAD with any type of PPC system, provided that both have a suitable I/O interface.

Direct access to the database of management systems via open program interfaces is an important requirement for keeping the data up to date. A heterogeneous arrangement of hardware may neccesitate the use of intermediate transfer files.

For the transfer of order related data and process planning technology data, the scope of design bill of material processors will in future also include the integrated processing of routing and cost calculation data generated in the CAD environment.

In order to put information about already existing similar parts at the designer's disposal, he must be given suitable access to the basic data of the PPC system. In this database the sets of part trees are usually arranged according to part numbers. Searching for certain part numbers, however, is impracticable for the designer, as precisely these numbers of similar parts are unknown to him.

By using a classifiction system as the central information system, the search method can be shaped to meet the designer's requirements. Not only DIN standard parts, but also individual components and parts can thus be classified and managed by means of their characteristic features if the single product components are gathered into groups with defined similarity, the so-called part families (in analogy with group technology). Characteristics sections are envisaged to enable the combination, delimination, selection, and representation of standardized as well as non-standard similar items. To conduct a search, the design engineer provides values or restricting parameters for several characteristics, which the information system then compares with the characteristics of the different part families (Fig. 6.2).

For the description of parts according to search terms or characteristics, DIN 4000 determines basic requirements. A characteristics section information system according to DIN 4000 can be used as a central information system, but it can also function as a means of coupling, especially for the data flow from CAD to PPC (Eigner *et al.* 1986).

The application of modern database systems for the storage of part tree data also enabling the formulation of complex search inquiries will in future diminish

Designation	Measuring unit	Selection criterion		
Nominal diameter	mm	A	=	"12"
Length	mm	& B	=<	"80"
Length of thread	mm	& C	>	"35"
Shape of head		& D	=	" Cylinder "
Height of head	mm			
Material		& F	=	"ST52"
Tolerance	%			

Fig. 6.2 Conducting a search according to characteristics.

the importance of classification (Steinmetz 1991). In the database no classes have to be created and encoded in advance. Instead the characteristics are included in the set of part trees as isolated data, showing their respective values. The characterisics to be traced can be combined to a complex search inquiry to the database system. Each search inquiry can thus be individually formulated.

Today, access to PPC data from the CAD system is mostly performed with the help of multifunctional graphic workstations. It is an organizational coupling; the user can switch from CAD to PPC system and back by means of a shift key and a channel controller. In many cases, the CAD computers are PCs which can be used as a conventional terminal of the PPC mainframe by terminal emulators. The design engineer is thus able to switch over from the CAD dialogue to the functions of the PPC system, and after switching back to continue the CAD dialogue. However, a direct transfer of PPC data to the CAD system cannot be accomplished.

As the normal PPC functions are often incapable of supplying relevant information to the design engineer in an appropriate manner, design information systems are being developed which can transform the desired information from the PPC data stock into a suitable form and put it at the disposal of the design engineer online (see for instance Nedeß *et al.* 1986).

6.2 A special coupling of CADIS and ISI

For the tool construction department of a manufacturing company a package of procedures was developed which implements a transfer of bills of material between the CAD system CADIS-2D by Siemens and the PPC system ISI (Industrial Control and Information System), also by Siemens (for the following cf. Scholz *et al.* 1988). The evaluation of drawings is performed with the help of the data structure evaluation module DASA by the Kraftwerks Union (KWU AG).

The data which are transferred from the CAD system to the PPC system come as a design bill of material which, by evaluating the CAD drawings, is generated automatically. This transfer is later supplemented by a coupling processor which performs the data transfer in the reverse direction from the PPC to the CAD system.

Within the general problem area of coupling and integration, the transfer of bills of material by a formatted file is only one of several alternatives. It is a relatively simple form of coupling which in spite of its separate implementation helps to keep the programming effort at the necessary minimum, and therefore in this case was chosen as a pragmatic solution. Due to the absence of standards, such a specific solution including the transfer of bills of material is often the only feasible way, especially where older PPC systems are involved.

6.2.1 Characteristic features of the tool construction area

The tool construction of a company influences the keeping of deadlines and the costs of manufacturing to a high degree. Tools and other resources required for manufacturing have to be made available at the appropriate point in time, and the costs involved in the construction of tools directly affect the overall costs of a product. Moreover, tool construction also indirectly affects manufacturing costs because of the tools' influence on processing times, auxiliary processing times and set-up times, as well as by the management of the tools within the tool cycle itself.

Alterations of target values by production control also change the requirements for tool construction: smaller lot sizes and shorter lead times require a higher number of tools and resources to be made available; the allocation and preparation times are shortened. As the manufacturing area has to meet ever stricter demands with regard to deadlines, the keeping of deadlines has consequently become one of the most prominent targets in the tools construction area.

Tool construction is mainly one-off jobbing production. Up to the present, for manufacturing and repairing complex tools a large number of adjustment operations have to be performed manually. These manual operations which account for a large share of the overall work are difficult to calculate and complicate precise planning of deadlines and costs.

Even though tool construction is a separate area within the company organization, it is not unconnected with the actual manufacturing area. Urgent repair and modification orders which have to be executed by the workshop without prior planning can severely disrupt the planned manufacturing processes. Often allocation methods which are oriented towards demand and long-term deadline planning are lacking. Inappropriate means of production control often cause delays, unplanned express orders, long throughput times, and a high, manual administrative effort (Ullmann *et al.* 1987). Throughput times can be shortened by automating and accelerating data exchange.

CADIS-2D is a two-dimensional system for interactive computer aided design. CADIS-2D is based on the software product SICAD-GR (Siemens CAD basic system). It follows the open system concept, and users can enlarge it by integrating their own commands. The stored image information can be evaluated via a standardized interface.

In the tool construction department of a user company, the number of designs generated with CADIS is steadily growing. For the generation and storage of design bills of materials, the interactive program TGS (Technical Basic Data Storage) is used. TGS can be regarded as the data registration part of the PPC system. Although TGS was not originally intended to be used in the tool construction area, the system is nevertheless being applied for the generation of bills of material.

The database of the PPC system ISI forms the basis of the TGS system. The ISI database uses the common PPC system data structure. Data derived from the drawings generated with CADIS have to be incorporated into this data structure. Root data are separated from structural data.

CADIS and ISI/TGS run on the Siemens operating system BS2000. In the framework of a systems analysis, a number of different possiblilities for coupling CADIS and ISI were studied (Scholz *et al.* 1987a):

- CADIS-PPS with an additional module and an interface to the production planning and control software IS by Siemens or another already installed PPC system (i.e. also ISI) was used only as a demonstrator model at the time of the study.

- An evaluation with the CADIS-AD function, a solution which in other departments of the user company had already been successfully applied with CADIS-sheet metal, would have involved a high effort and would therefore have been very slow due to the peculiar conditions of the department concerned.
- The enlargement of CADIS with completely customized FORTRAN evaluation procedures would have been quite labour-intensive.
- The data structure evaluation module DASA by KWU is an enlargement which already provides certain evaluation functions: a supplementary package to the Siemens CAD system CADIS-2D which can evaluate drawings and plans generated with SICAD, CADIS or other SICAD based enlarged systems. Information and data established by the DASA procedures can be transferred to different output media: screen, printer, SICAD system variables, or BS2000 file.

With the help of the DASA module, data from the CAD application can be put at the disposal of other programming systems on a Siemens 7.5xx mainframe, in the present case the ISI database. DASA can be applied interactively, in batch, or integrated into SICAD procedures.

The work methods of design engineers using CADIS-2D and TGS for drawing and bill of material generation vary. Data not included in the CADIS drawing are first noted manually and later interactively handed over to TGS. These data are statements about standard parts and those purchased parts which do not have to undergo further processing.

Also statements already included in the drawing have to be input again in the TGS dialogue. These can be up to 700 items of the bill of material for each design, a task which takes the design engineer about one day to perform. During this time he does not do any creative work. Until now only semi-finished materials, i.e. standard parts or purchased parts which need further processing (manufacturing parts) are included in the drawing. Bill of material head data needed in the TGS dialogue are located in the text section of the drawing.

The implementation achieved automated transfer of data included in the drawing to the TGS. For drawing evaluation it uses the evaluation commands of DASA. At the same time, a number of CADIS commands are also used, especially for the selection of symbols to be evaluated and for plausibility checks in advance of the evaluation process. These commands are combined into command sequences in procedures. Error-free evaluations of CADIS drawings are stored in a transfer file and are thus cleared for further processing by the TGS module.

The procedure package is a converter which collects, converts, and clears for further processing data from CADIS drawings. The coupling of CADIS-2D and TGS itself is thus achieved by the generated transfer file, an intermediate type of a bill of material which on the one hand is adapted to the features for automatic drawing evaluation of DASA, and on the other hand takes into account the requirements of the subsequent processing steps. It is later transformed again by a coupling module in such a way that it can be received by the ISI database (Fig. 6.3).

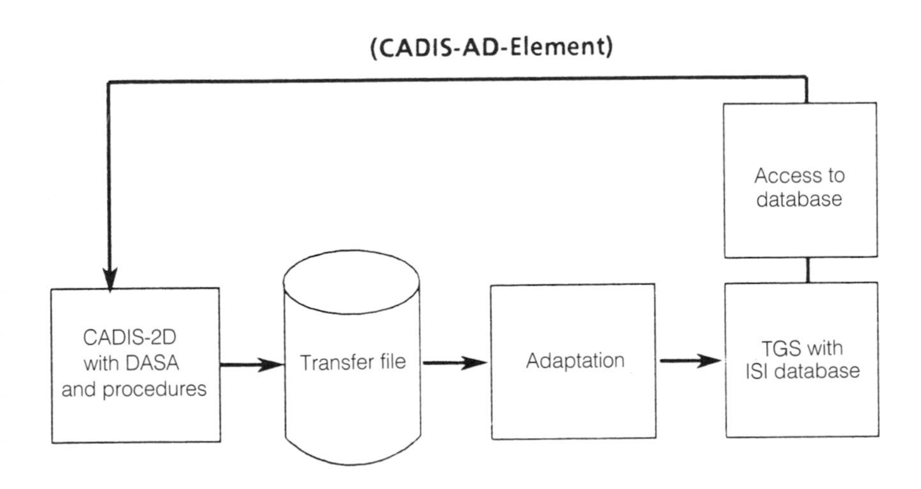

Fig. 6.3 Coupling of CADIS-2D and TGS with transfer file and adaptation module (Scholz *et al.* 1988).

6.2.2 Organizational aspects of the specific solution

With the help of the procedure package, all individual images stored in an image library can be evaluated for the generation of bills of material by one command. As a precondition, specific symbols which can be assessed by the procedure package have to be used to mark the drawing's head section and the head sections of parts drawings. Thus new symbols were defined for the head section of drawings, parts drawings, and their corresponding head section attachments which are better adapted to the later shape of the bill of material than those previously used (Fig. 6.4).

The volume of useful data is still limited to the statements contained in the drawing's head section and data about parts which are to be manufactured. Therefore statements about standard parts and purchased parts must still be input interactively with the TGS. The functional range of the package can be enlarged by the evaluation of symbols which can also record and represent these data.

When all the parts of a construction drawing are available, i.e. the main drawing with its following pages and the various part drawings with their following pages, the automated evaluation can begin.

The call to the evaluation program is performed in the command mode of the operating system BS2000. Afterwards the system switches to the program mode and asks the user for the name of the image library to be evaluated. This name is usually the identification number of a drawing. Now the automated evaluation starts: the program activates CADIS-2D, compiles a list with the names of all single images contained in the image library, and then generates a special evaluation procedure for this library. This procedure is automatically started by the program. Now one by one each of the images is loaded into the system and evaluated by DASA.

The program interprets all the individual images within the target library as parts of a complete composition drawing. Therefore all the single CADIS images of a drawing (main drawings and part drawings) - and only those of one drawing - have to be included in the library.

Before the evaluation procedure is started, plausibility checks on some entries in the symbols are performed. Single images without a drawing head section are not evaluated as essential statements for the evaluation of part drawings must also be derived from the head section. The head section of the overall drawing is explicitly evaluated only as the single image which is marked as the first page of the main drawing.

Modified drawings cannot be automatically evaluated: for the execution of modifications in its database, TGS requires the input of the modified data, and of those data only. The programming effort involved in a program package which would enable such a differentiation in the CADIS image data files would by far exceed its benefits. Bills of material for modified CADIS image data files therefore have to be generated interactively with the TGS. The program examines a reference indicating a modification in the status section within the drawing's head section.

Symbol TZ.N

-----	-------------------	----	----	-------	--	-
Quantity	Designation	Serial number	Code number	Item number		
	Tolerance acc. to DIN 7168 rough	----	----	96-----	--	-

Symbole TZ.ST.A

----	------------------	----	----	----------	----

----	------------------	----	----	----------	----

----	----------------	----	----	---------	----
Quantity	Designation	Position	Code number	Item number	Remarks
	Tolerance acc. to DIN 7168 rough	----	----	96-------	----

Symbol TZ.ST

Symbol K02

---	--	--	--	Inspection number	Designation	Serial number	Code number	Item number	Remarks

Scale of original

Surface

State	Message	Date	Name	
				Date
				Person in charge
				Checked
				Standard
				Phone number

AG

Neighbour: yes/no

Replacement for Replaced by

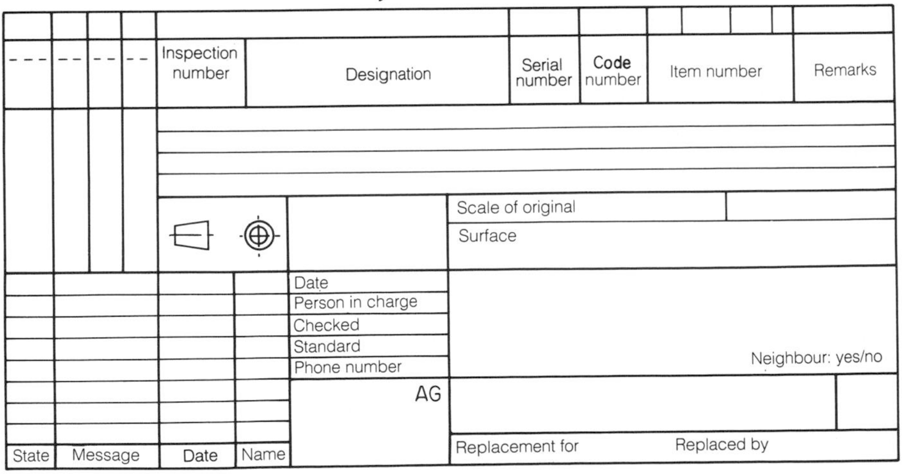

Fig. 6.4 Old and new head sections of drawings (Scholz *et al.* 1988).

The position numbers in the head sections of the part drawings and their symbols are also examined: identical position numbers for different parts are rejected, and so are inadmissable entries.

During the evaluation, the program continuously informs the user about the work progress. Most of the data the program gathers from the drawings is shown on screen. Moreover the program indicates the state of evaluation and which symbols were found in the processed image and which were not.

Once all the single images of the image library/composition drawing have been completely evaluated, a second check is conducted on the position numbers. Only now the statements contained in different images can also be compared with each other. If the same position number is found twice in the evaluation file, the program terminates with an error message. Otherwise the current evaluation is cleared for transfer to the TGS system. Then the evaluation of the composition drawing is completed, and the operating system returns to the command mode.

If the program terminates with an error message, the evaluation conducted up to this point is void. The errors detected by the program have to be corrected and the evaluation procedure has to be performed again from the beginning.

The user does not have to pay any attention to the co-operation with the TGS module. The transfer of the generated file to the ISI database is periodically conducted at night. This determines the maximum span of time elapsing until the automatically evaluated information is available in the ISI database. Afterwards, missing information can be added interactively with the TGS. This information includes data about standard parts and purchased parts. If the functional range of the software package for the automatic evaluation of CADIS drawings could be enlarged in such a way that these data could also be detected, the dialogue with the TGS module would generally be unnecessary.

It would also be possible to insert data from the ISI database directly into the CADIS drawing. This is especially interesting in the case of designation texts of standard parts which can then automatically be recalled by their stock item number and be inserted into the corresponding symbol table.

6.2.3 Technical aspects of the specific solution

The software package consists of three types of procedures which call up each other:

- BS2000 procedures for the operating system level (DO procedures),
- a procedure for the BS2000 file processor EDT (INPUT procedure), and

- CADIS/DASA user procedures for the user program level of the CAD system CADIS-2D (DASA procedures),

The procedures operate with the following software products:

- SICAD-GR version 4.3B1,
- CADIS-2D version 1.3.B, and
- DASA version 1.3B.

CADIS is an enlargement of the basic graphics system SICAD-GR. SICAD-GR was implemented in FORTRAN 77 and Assembler and forms the basis for almost all CAD products by SIEMENS.

Data structure management plays a pivotal role in the SICAD system. The model structure of a drawing is stored in a model data structure. CAD functions for insertion, manipulation, output, or deletion are performed within this data structure. The necessary function routines are accessed by commands.

All elements of the model data structure (e.g. circle (CR), line (LI), symbol (SY) or text parameter (TP)) have an identical structure: the element header contains all important management information, elements of the same type are linked with each other by a base chain, elements of different types (a 'master' and the 'details' it consists of) are linked into a ring structure by the structure chain, in the parameter section geometrical and non-geometrical data are stored, and the descriptor section is often used as an addition to the parameter section. However, parameters and descriptors may be left out.

Via the structure chain, the drawings are further and further detailed, starting with the uppermost 'master' (Fig. 6.5). For the evaluation of drawings, these structure chains have to be searched for header symbols and the contents of the corresponding text parameters must be examined.

DASA is a CADIS-2D system enlargement for special applications which was developed by KWU and provides functions for evaluating the model data structure. The functions of DASA were implemented in FORTRAN 77; they were structured and programmed according to the same principle as those of SICAD. When DASA is incorporated into SICAD or its enlargements, several DASA and CADIS commands, which with the help of the SICAD procedure technology are written in a sequence, can be combined into user-defined procedures which can be started by a dialogue command.

In the overall procedural sequence, the DASA user package is activated after the start of CADIS-2D with DASA. It begins with a procedure which loads all single images within an image library one after the other, and for each image calls the real main procedure for single image evaluation, which in turn activates the other

DASA procedures. The search for drawing head section symbols is conducted with CADIS selection commands. DASA functions carry out the evaluation of text parameters and attributes. Figures 6.6, 6.7, and 6.8 show an examplary drawing with the transfer file which was derived from it.

The administrative/organizational application system TGS (Technical Basic Data Storage) was implemented in the programming language COBOL and - like SICAD - runs on a Siemens system 7500 with the operating system BS2000.

The database of TGS, the ISI database, has the conventional PPC system data structure. The ISI system operates with a part master file containing part related data as well as item data for shop order execution, a part structure file which links the data records of the part master file, and an index file which functions as a table of contents of the ISI files. In order to make data which have been stored by the procedure package in a transfer file available to the ISI system, they have to be incorporated into this database.

The data format within the transfer file which is required for including data into ISI is split up into root data records, structure data records, and comment data records. Structure data records are generated for each part used in a design. If there are some comments on it, an additional comment data record is necessary. During the transfer process, new root records have to be created for the drawing number and each of its versions and furthermore for all data records of parts not on stock which are used in the design.

The transfer file generated with DASA only differentiates between root records for each drawing number (type *01*) and structure data records (type *02*) and therefore must be adapted. From each data record of the record type *01* two root data records of the required format are generated: one for the assembly item number and one for the first variety of the assembly. The contents of both records are largely identical. From each data record of the type *02*, a structure data record of the required format is generated. If this data record contains comment text, an additional comment data record must be generated. In addition to that, for parts which are not on stock a root data record must be generated.

As a high degree of accuracy is essential for the data of a PPC system, a number of data records of the transfer file have to undergo further plausibility tests before they are admitted into the database.

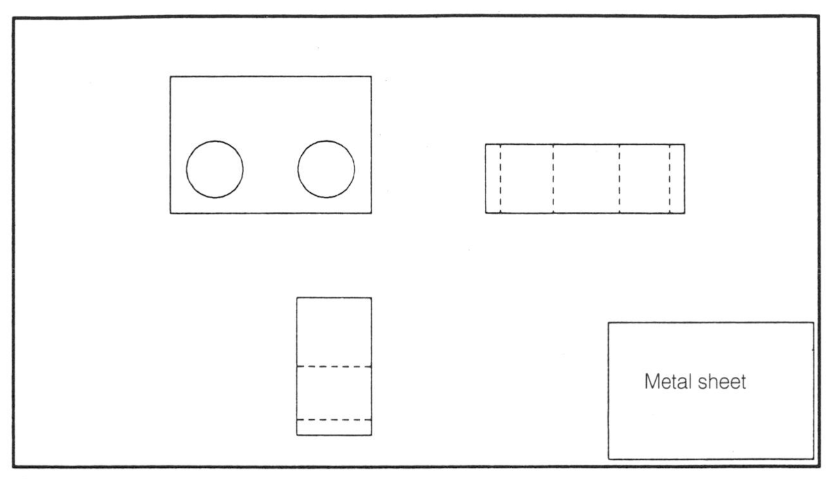

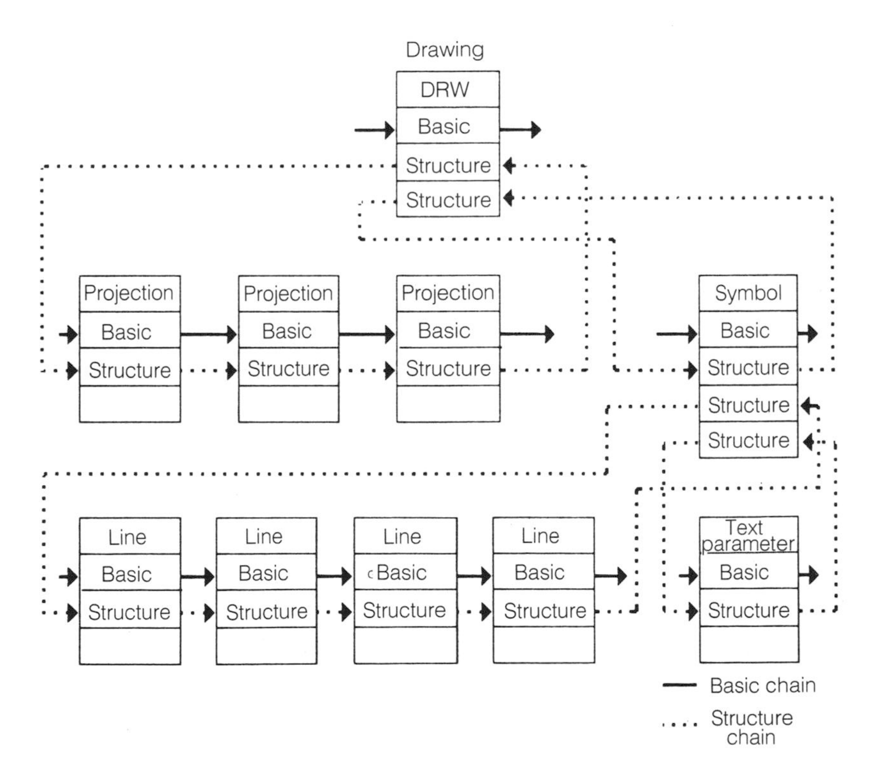

Fig. 6.5 Data structure of the drawing 'metal sheet' (Scholz *et al.* 1988).

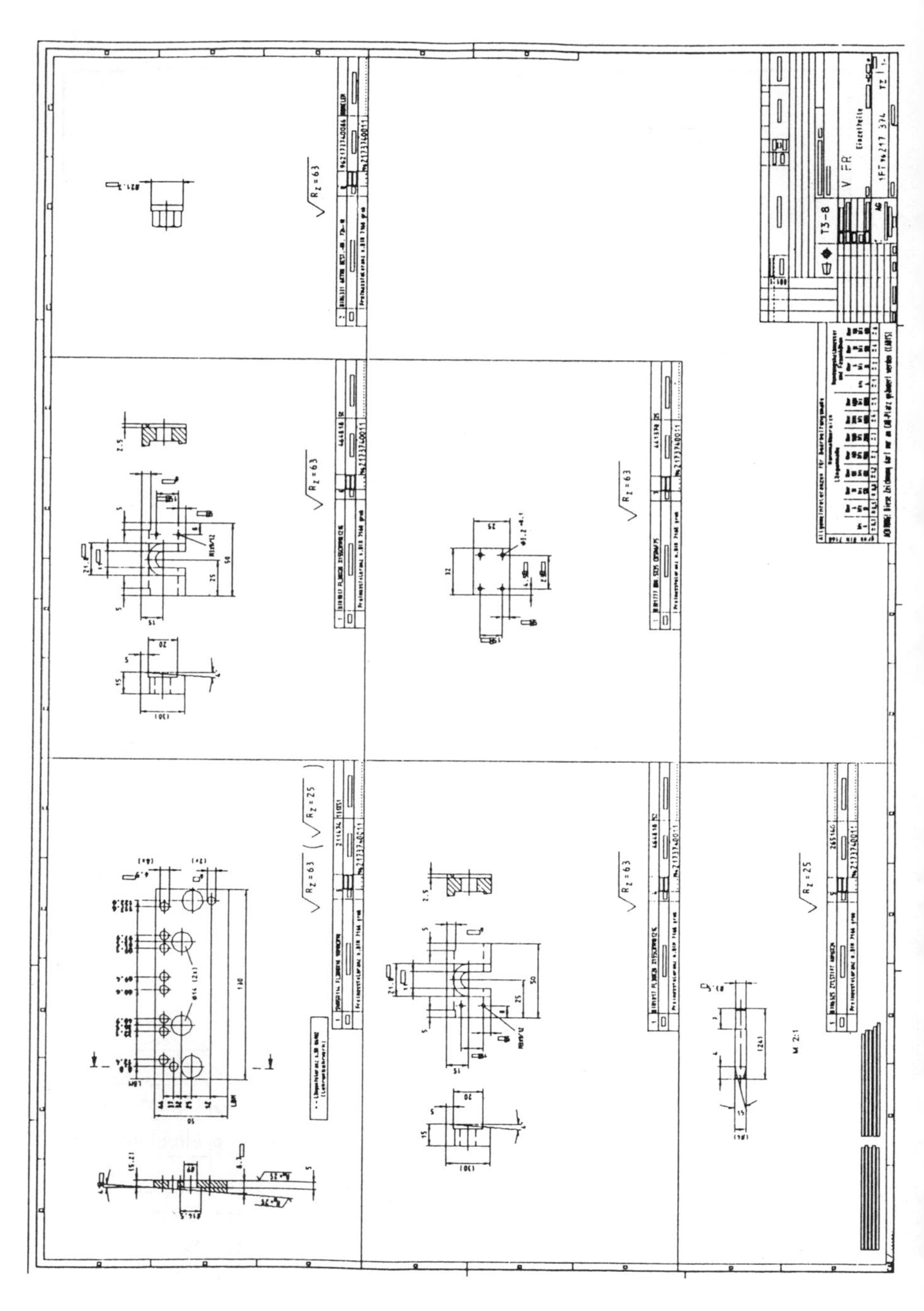

Fig. 6.6 Component drawing (Scholz *et al.* 1988).

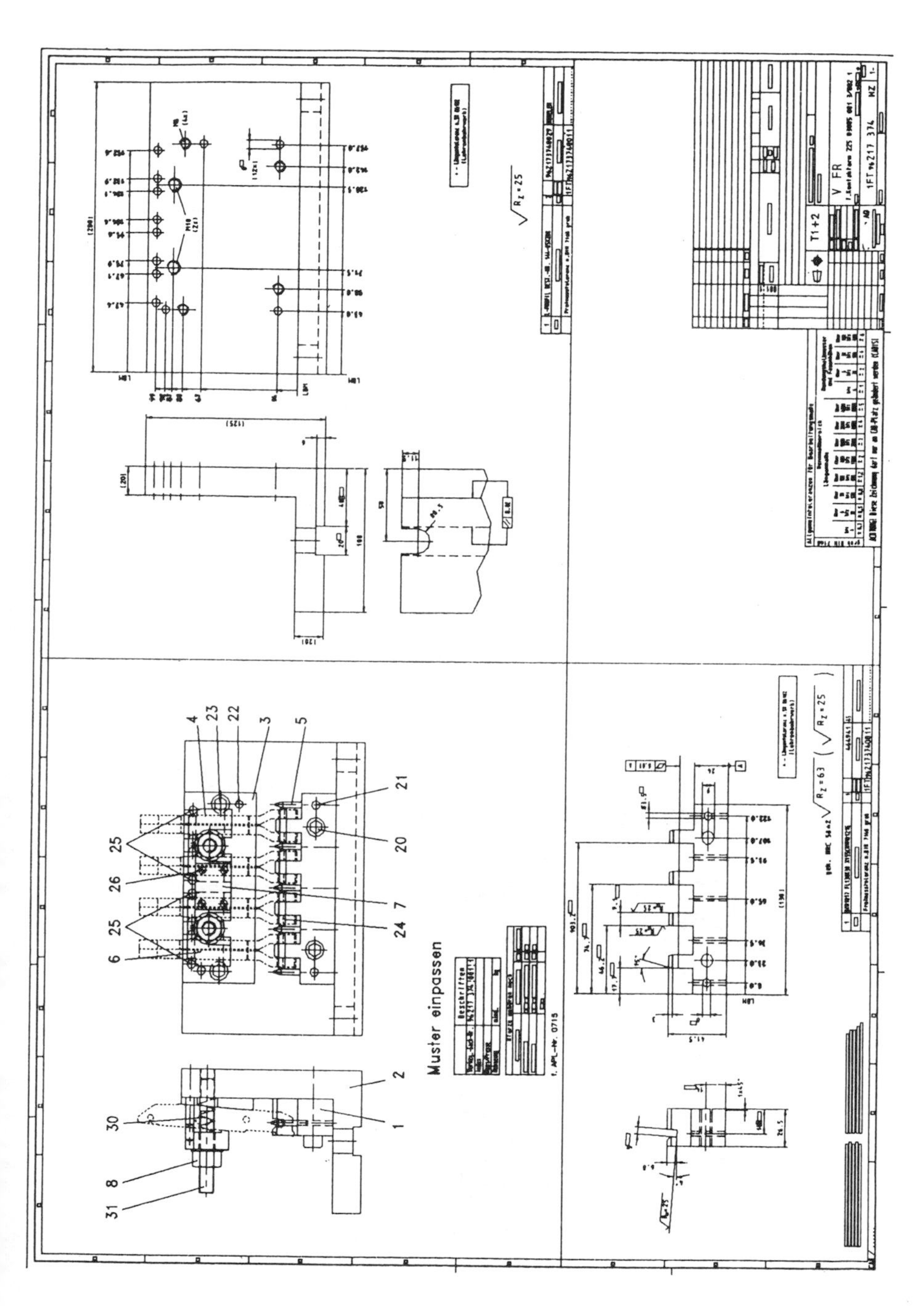

Fig. 6.7 Composition drawing (Scholz *et al.* 1988).

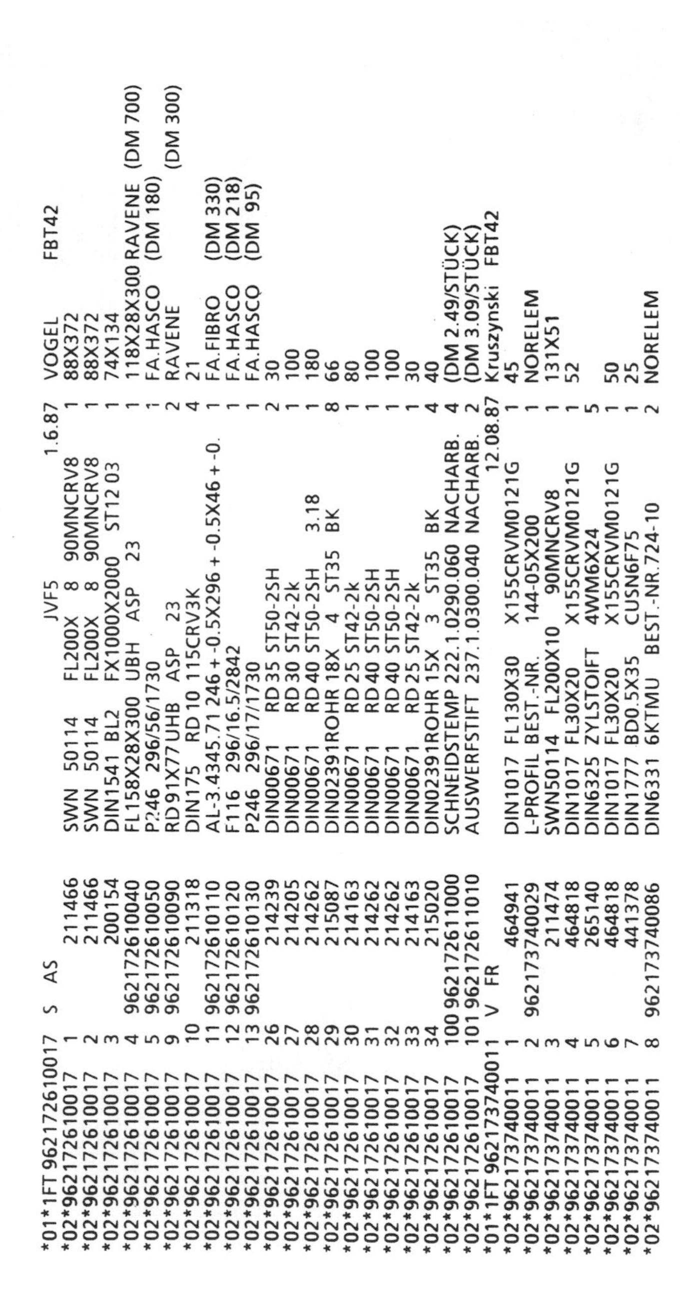

```
*01*1FT 962172610017  S   AS                                    JVF5                 1.6.87  VOGEL        FBT42
*02*962172610017   1           211466    SWN  50114    FL200X   8   90MNCRV8        1   88X372
*02*962172610017   2           211466    SWN  50114    FL200X   8   90MNCRV8        1   88X372
*02*962172610017   3           200154    DIN1541  BL2    FX1000X2000     ST12 03      1   74X134
*02*962172610017   4   962172610040     FL158X28X300  UBH    ASP    23                 1   118X28X300 RAVENE  (DM 700)
*02*962172610017   5   962172610050     P?46   296/56/1730                             1   FA.HASCO     (DM 180)
*02*962172610017   9   962172610090     RD 91X77 UHB    ASP    23                      2   RAVENE                 (DM 300)
*02*962172610017   10          211318    DIN175    RD 10  115CRV3K                     4   21
*02*962172610017   11 962172610110      AL-3.4345.71 246 + -0.5X296 + -0.5X46 + -0.   1   FA.FIBRO     (DM 330)
*02*962172610017   12 962172610120      F116   296/16.5/2842                           1   FA.HASCO     (DM 218)
*02*962172610017   13 962172610130      P246   296/17/1730                             1   FA.HASCO     (DM  95)
*02*962172610017   26          214239    DIN00671     RD 35  ST50-2SH                  2   30
*02*962172610017   27          214205    DIN00671     RD 30  ST42-2k                   1   100
*02*962172610017   28          214262    DIN00671     RD 40  ST50-2SH     3.18         1   180
*02*962172610017   29          215087    DIN02391ROHR 18X    4   ST35   BK             8   66
*02*962172610017   30          214163    DIN00671     RD 25  ST42-2k                   1   80
*02*962172610017   31          214262    DIN00671     RD 40  ST50-2SH                  1   100
*02*962172610017   32          214262    DIN00671     RD 40  ST50-2SH                  1   100
*02*962172610017   33          214163    DIN00671     RD 25  ST42-2k                   1   30
*02*962172610017   34          215020    DIN02391ROHR 15X    3   ST35   BK             4   40
*02*962172610017   100 962172611000     SCHNEIDSTEMP 222.1.0290.060  NACHARB.         4   (DM 2.49/STÜCK)
*02*962172610017   101 962172611010     AUSWERFSTIFT 237.1.0300.040  NACHARB.         2   (DM 3.09/STÜCK)
*01*1FT 962173740011  V   FR                                                 12.08.87  Kruszynski   FBT42
*02*962173740011   1           464941    DIN1017  FL130X30     X155CRVM0121G          1   45
*02*962173740011   2   962173740029     L-PROFIL  BEST.-NR.    144-05X200              1   NORELEM
*02*962173740011   3           211474    SWN50114   FL200X10     90MNCRV8              1   131X51
*02*962173740011   4           464818    DIN1017  FL30X20      X155CRVM0121G          1   52
*02*962173740011   5           265140    DIN6325  ZYLSTOIFT    4WM6X24                 5
*02*962173740011   6           464818    DIN1017  FL30X20      X155CRVM0121G          1   50
*02*962173740011   7           441378    DIN1777  BD0.5X35     CUSN6F75                1   25
*02*962173740011   8   962173740086     DIN6331  6KTMU    BEST.-NR.724-10              2   NORELEM
```

Fig. 6.8 Transfer file (Scholz *et al.* 1988).

In order to enable the direct transfer of data from the ISI database into the CADIS design drawing during the design process, the symbols which are to absorb these data are defined as so-called AD elements. The AD function 'user specific data management' of CADIS version 1.3B allows a comparatively simple input of data and their entry into the text sections of the symbols.

With the help of a supplementary program enabling access by CADIS to the ISI database for this data input according to the stock item number via the ISI index file, ISI data then can be directly transferred to the text sections of the symbols defined as AD elements. These data transferred by the system do not have to be provided by the designer via the keyboard.

6.3 Summary

Couplings of PPC systems and CA applications have mainly been implemented in the CAD and - less often - in the CAPP area. In contrast to the CA application area, no interface standards yet exist.

As an interface for the coupling of CAD and PPC, the design bill of material generated with CAD and transformed into a transfer file plays the dominant role.

In case of a coupling of PPC to CAD, the design engineer should be provided with PPC data interactively via an information system. Characteristics sections according to DIN 4000 can form a suitable aid for the detection of similarities. An information system of this kind can make a positive contribution to decision making in the design area which to a considerable degree influences costs. Today, access to PPC data from the CAD system is often achieved with the help of multifunctional workstations via terminal emulation.

With an order interface, the carrying-out of design operations is integrated into the PPC order execution control. Numerous couplings applied in practice are based on simple methods of data transfer.

The coupling of CAD and PPC systems can be performed with a relatively low programming effort even if older software is used (in case of the previously described evaluation procedures of the special coupling CADIS - ISI about one man month), by using a formatted file (design bill of material) and a coupling processor. Although the solution described above is a specific one, it is based on software for drawing evaluation available on the market. Some restrictions have to be tolerated (e.g. no evaluation possible in case of alterations). The implemented coupling does not meet the demands which ideally a coupling of the CIM components CAD and PPC should fulfil. It nevertheless shows that even a relatively small-scale solution can have beneficial effects and can be meaningful, even when older software is used.

Chapter 7

Comprehensive concepts for coupling in CIM

There are numerous possibilities for implementing CIM on the basis of the concepts and requirements of the applied systems described above. The coupling of different software systems fulfilling different tasks cannot be avoided when implementing CIM systems.

However, when applying the described methods for coupling and integration, there is always the danger of creating new, though larger, isolated solutions in which some of the hitherto isolated applications have been coupled but interfaces to other adjacent systems are being disregarded.

The missing link is a factory or company wide administration which co-ordinates and couples all control and data flows, thereby establishing the precondition for functional integration.

In contrast to the CIM concepts published in Germany, this issue is reflected in the CIM wheel proposed by the Computer and Automated Systems Association (CASA) of the Society of Manufacturing Engineers (SME) (see Chapter 1).

The hub of the wheel represents information and communication management for the coupling of the single application areas which are using a common database. A common database alone is thus insufficient for attaining a really comprehensive concept for coupling.

Some approaches aiming at a comprehensive information and communication management for the coupling of CIM applications (which by the way closely resemble each other) will be introduced in this last chapter. Only tentative steps - if any - have been taken towards an implementation of these concepts.

7.1 Action oriented data processing

Action oriented data processing (AODP) (for the following paragraphs cf. Mertens *et al.* 1986, Hofmann 1988) is a continuation of batch oriented but nevertheless integrated data processing which in practice has been implemented in classic, modularized PPC systems, enlarged by interactive applications.

AODP is beneficial for the integration mainly of process chains which show a high degree of labour division, or which are heterogeneous with regard to their data processing organization. Information flows, deadlines, etc., are determined and co-ordinated by central data processing. Depending on the task to be executed, room for human decision making in certain cases has to be left. AODP therefore can be described as a system of integrated dialogues.

Mertens and Hofmann describe AODP as a combination of the following ideas, concepts, and technical systems (Mertens *et al.* 1986):

- central control of labour divided processes by dispatchers, command stations, etc., which show a good degree of information and have the capability of initiating and supervising sequences of actions of medium complexity; these sequences do not always follow the same routine path and therefore cannot be fully automated (yet);
- man-machine dialogue;
- integrated data processing;
- electronic mail and office communication;
- administration largely independent of documents and records;
- critical path method;
- generalization of PPC systems and their elements to common administrative tasks; and
- trigger files in data management systems.

A system based on the AODP concept is shown in Fig. 7.1.

The function oriented application system consists of a range of various types of functional applications. Existing applications have to be adapted with regard to registration, generation, distribution, signaling, and processing of messages to and by the message handling and distribution system.

Within the message handling and distribution system, different types of messages can be identified according to the type of addressee.

Action messsages are addressed to certain persons or offices as a reminder of actions due; the action messages are left in the mailbox of the person or office in charge. This is a standard function of an electronic mail system.

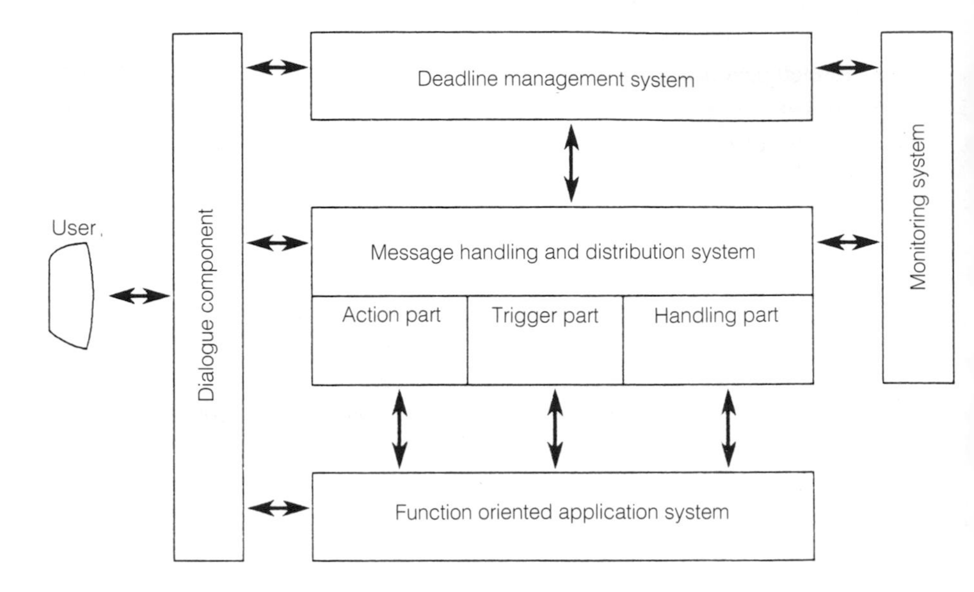

Fig. 7.1 Components of an AODP system (Mertens *et al.* 1986).

Trigger messages activate programs of the function oriented application system or the monitoring system, partly inclusive of data transfer. The trigger concept is implemented in many integrated PPC systems (e.g. in ILAS by ADV/ORGA).

Handling messages trigger physical, computer controlled events, i.e. transport, storage, manufacturing and assembly events, or they transfer data to the corresponding executing systems.

The required procedural parameters of trigger and handling messages are managed via corresponding tables. These messages are placed into an intermediate buffer by the triggering program. The buffer is constantly accessed by a call controller which automatically activates the required programs or physical processes when the appropriate capacity and other secondary conditions are given.

The deadline management system mainly provides functions for deadline management, planning, and issuing (functions of this type are offered for instance by the office information system ALL-IN-1 by Digital Equipment). The deadlines are supervised by the monitoring system which again with the help of the message handling and distribution system controls the current work-load on all partial systems (especially on the functional applications) and occasionally triggers due actions. In this context, network plan technology plays an important role.

The dialogue component is usually menu-controlled and provides a unified user interface. It handles the man-machine dialogue.

Major application systems in the manufacturing area which are based on the AODP concept are the PPC system COPICS (Communication Oriented Production Information and Control System) by IBM, and the system CIMOS (Computer Integrated Modular Organization Software) by MTU. A system covering all CIM areas does not exist.

According to Mertens and Hofmann, AODP aims at attaining the following goals:

- completeness/complexity handling capability;
- guarantee of complete execution even of complex process chains with a high degree of labour division;
- strict co-ordination: actions on critical paths should be assigned a special priority and specific supervision; lead times should be as short as possible;
- good utilization of capacities: the single places of action should not be overloaded; there should be the possiblity of taking capacity balancing measures;
- optimal division of labour between man and machine;
- less use of paper as an information carrier.

(Mertens *et al.* 1986)

7.2 CIM Handler/INMAS

The concept of a CIM handler (for the following cf. for example Scheer *et al.* 1990, Gröner *et al.* 1987) was developed at the Institut für Wirtschaftsinformatik in Saarbrücken, Germany. According to this concept, existing isolated solutions are integrated by additional software packages. In industry, a wide range of CIM components have already been installed, some of which are quite mature and cannot be replaced easily. As a completely new implementation of all CIM components would be very expensive and time-consuming, integration via a CIM handler or an individually configurable interface (INMAS - Interface Management System) has been proposed.

The CIM handler has a modular structure and, as a central system, controls the information exchange between the single components. It ensures that the information about modifications of one of the connected systems are transferred to every other system for which these modifications are relevant. The consistency of the operational database is thus maintained without any interaction by the user.

The emergence of new isolated solutions on a higher level by coupling single components with each other is avoided. Such a CIM system can be configured individually and with a high degree of flexibility (the components to be integrated can be selected on a case-by-case basis) provided that the handler itself is adaptable to the same degree.

The kernel functional range was given a modular structure for development and operation. The following functions were included:

- A terminal emulator is a precondition for establishing bi-directional communication between the geometrical/technological and the administrative computer environment. It must ensure the communication of each application with other systems without the necessity of terminating the application itself.
- Data exchange transactions ensure a complete distribution of newly generated data among all the connected systems. When data are generated or modified by one system, these generated or modified data must then be transferred automatically to all other systems concerned, whereby in some cases problems of co-ordination (which data are to be transferred when and to which system) must be solved interactively.
- Information transactions enable predefined or freely formulated inquiries to the database from all connected systems with access to all non-protected data.
- The system-wide status handling module registers the execution state of every data exchange transaction.
- The trigger concept enables the automatic call-up and handling of a wide range of different activities due to certain states.
- Data dictionary functions, i.e. the automatic conversion of different data formats when systems exchange data with each other can - if required - be performed with the help of the data dictionary.
- Customizing enables the user to configure his CIM system freely during installation or adaptation. Permitted transactions are stored in a transaction database. These transactions can be individually combined into various transaction sequences.

Data transfer is performed via transfer files, the contents of which are supplemented by information about all connected systems and checked with regard to plausibility before they are split up for transfer to the databases of the different components. Each target system disposes of a set of operation reserve files which it gradually processes in order to finally integrate the data.

The CIM handler can reduce the high effort with regard to time, personnel, and capital involved in the whole process of conceptualizing an integrated system with all functions operating without faults, up to its introduction and the training of its users.

Software vendors could offer a CIM handler as a modular standardized program which, via a shell, can connect several systems offered by the company with each other. For certain functions, users may also use components of other vendors if these better suit their requirements.

The CIM handler is structured according to the shell model: state management as the kernel is superposed by the transaction sequence definitions which in turn access the transactions. The link to the connected systems is established by a neutral data shell.

As far as the CIM handler cannot influence it, the user interface of the different components remains unaltered. Although a unified user interface for all CIM components should be demanded, the CIM handler concept described here can only support to a limited extent this request as it mainly addresses the problem of connecting isolated solutions which themselve remain unchanged. A unified user interface for CIM applications is the subject of the partial project 'Man Machine Interface' within the ESPRIT project No. 2527 'CIDAM', of which also INMAS is a partial project.

The modules to be connected by the CIM handler must fulfil the following requirement:

- A hardware connection with the corresponding basic functions (file transfer and terminal emulation) is a precondition.
- The software systems to be connected must as a rule have user exits in order to enable these systems to co-operate with the CIM handler software.
- The structure of the specific databases must be known. Also other systems must be able to access them via the CIM handler.

Scheer also demands the implementation of unified system parts such as for instance a relational database system. This demand can be put into practice by employing advanced database concepts which can be implemented on several levels of the hardware hierarchy (microcomputer, minicomputer, mainframe).

Taking the long view, research and development will have to address the development of a knowledge-based CIM handler. The following functions will have to be included in such a CIM handler (Krallmann 1986):

- sequencing, organization, control, and diagnosis of systems in a network in case of failure of a node;
- control of security and restart procedures;

- planning, control and supervision of manufacturing processes on different enterprise levels.

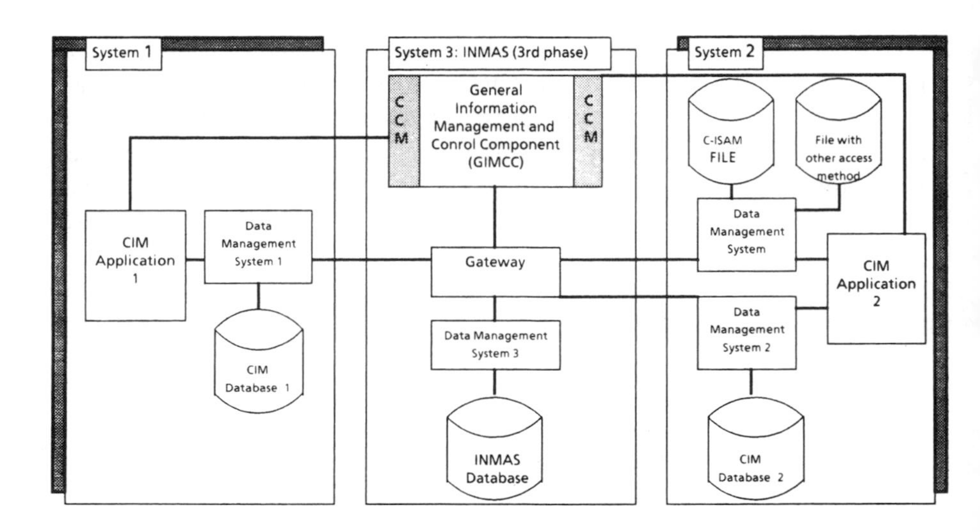

Fig. 7.2: Components of INMAS (Scheer *et al.* 1990).

The CIM handler concept forms the basis for the individually configurable interface INMAS (Interface Management System) which is the target of research to be conducted in the framework of the ESPRIT II project CIDAM during the coming three or four years (cf. Scheer *et al.* 1990). The pre-eminent kernel of this interface is a neutral data structure. With regard to data integration, the present conceptual description of INMAS closely resembles the results of the existing concept of STEP (cf. Chapter 3). Concerning process integration, INMAS to a large degree will be based on action oriented data processing. The draft interface LIFE (Logistics Interface for Management Environment) to some extent shows a similar approach (cf. Göbel *et al.* 1990), although it merely addresses the coupling of the functions PPC and CAM.

7.3 The CASH concept

The CASH concept (computer aided simulation and information handling) by Siemens/KWU in principle shows the same approach as the CIM handler concept. It also pre-supposes an already existing organizational CIM structure with isolated data processing solutions and several central procedures. Particular attention was given to the fact that an enterprise structure does already exist which has generated its own data processing environments in the technical, manufacturing technological, and the administrative areas.

The main characteristics of the CASH concept can be outlined as follows:

- For action control, a Petri net is used instead of a central status file or a status database. This enables the representation and control of chronologically ordered and interlocking processes. Job variables (BS 2000) serve as central control elements (semaphores).
- CASH consists of a range of modular components each serving two or three procedures; this leads to a higher degree of fail-safety than a central module.
- Expert systems are not part of CASH but can only be applied within the single procedures. For the supervision and control of the Petri net status, common supports such as guided dialogue are available.
- For all planning components, CASH provides simulation-and-modelling; i.e. a 'let's-try-and-see-what-happens' function is *a priori* provided.
- CASH includes an intercommunicative message component, a so-called mailing system, with message exchange 'procedure-to-procedure' (Petri net), 'procedure-to-man' (interactively via mailbox), and 'man-to-procedure' (Petri net).

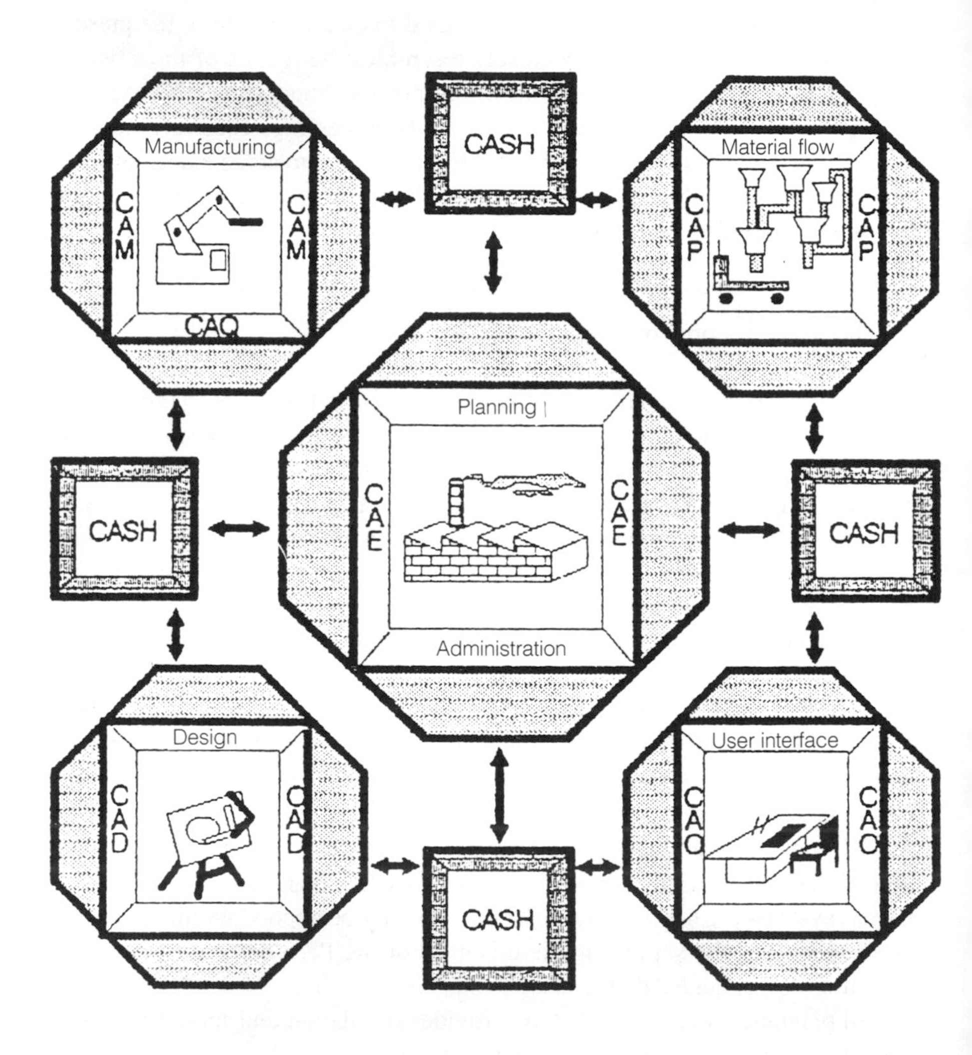

Fig. 7.3: CASH concept (Sedlaczek 1988).

7.4 Blackboard architecture

In the following paragraphs, the blackboard architecture as an example of a coupling concept originating from the artificial intelligence research domain is described (cf. Nii 1986, Hayes-Roth 1985) (other concepts as for example the contract net model (Davies *et al.* 1983) also exist). First applications addressed the speech recognition area.

A software system with this architecture is expected to be able to more or less simultaneously organize or process several voluminous, independent, and heterogeneous knowledge bases (Harmon *et al.* 1985).

In the current context this architecture assumes special relevance if an expert system is applied as the information and communication manager which possibly even serves further functional expert systems.

The expert system as information and communication manager thus processes several voluminous specific knowledge bases. These again transfer information relevant for the solution of a given problem to a shared, global area, the blackboard. Depending on the state of the problem solving process, it then contains relevant information originating from several knowledge bases concerning the solution of a problem.

The knowledge bases produce information which lead step by step to the solution of a problem. The entire communication and interaction between the knowledge bases is conducted exclusively via the blackboard. Dependending on the blackboard's state, the knowledge bases control their information transfer to it by a self-activating procedure. This concept can be compared with the metaplan method, where human participants reacting to the information which is written on the blackboard ask leave to speak and contribute further information to solving a problem.

The basic model described here is only a concept. For implementation within a real software system it has been further specified by Nii (1986). The refinements or the deviations from the basic model were found by studying systems which basically follow the blackboard architecture. According to these studies, a blackboard system has the following features.

- Knowledge bases

The knowledge relevant for the solution of a problem is split up into separate and independent knowledge bases. Each knowledge base aims at contributing information for the solution of the problem. The knowledge bases represent procedures, rules, or logic statements. The knowledge bases only alter the blackboard or the control data (which may be stored on the blackboard). The blackboard is only altered by the knowledge bases. Each knowledge base is responsible for knowing when it can contribute to the solution of the problem.

- Blackboard

The current state of the problem solution procedure is stored in the blackboard area. The blackboard has the task of supplying the several knowledge bases with information concerning the shared data relevant for the solution of a problem and the status of the problem solving process. The knowledge bases indirectly communicate with each other via the blackboard. The blackboard contains the objects of the solution space. These objects may be input data, partial solutions, alternative solutions, and final solutions, and possibly also control data. The objects are hierarchically ordered according to analytical levels. The objects and their properties define the solution space. The relationships between the objects are determined.

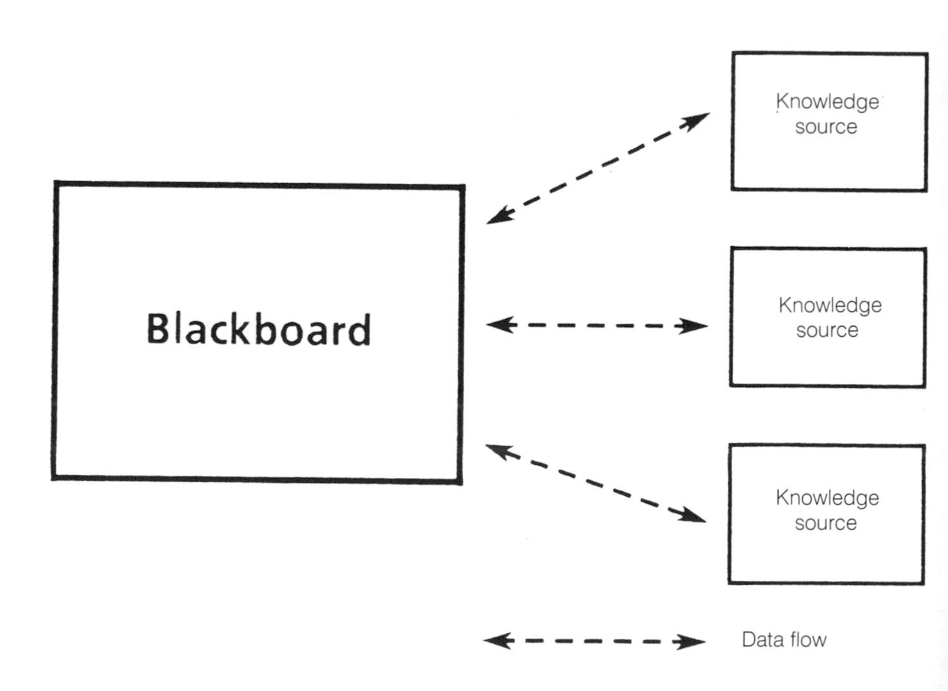

Fig. 7.4 Basic model of the blackboard architecture (with reference to Nii 1986).

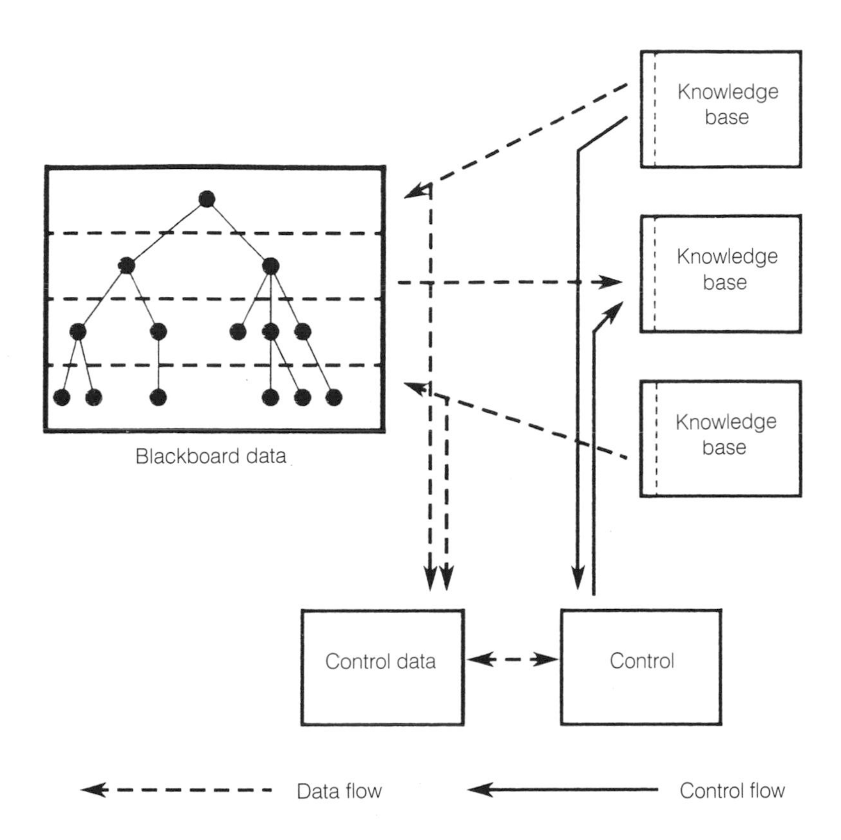

Fig. 7.5 Blackboard system (according to Nii 1986).

- Control

  In addition to the basic model a control module is introduced which selects and induces the execution of the next knowledge base, depending on the state of the blackboard and the control data.

Up to the present, the blackboard architecture has not been applied to information and communication management. However, probably the most widely known system for production planning and control, ISIS resp. OPIS, was organized according to blackboard architecture principles. The system organization contains several knowledge bases for representing alternative planning strategies and heuristic procedures for sequence control (Fox *et al.* 1984, Smith *et al.* 1986).

The blackboard architecture constitutes a special form of communication which could again be applied for AODP.

Craig describes an implementation of the blackboard approach for air traffic control purposes (Craig 1989).

7.5 Integrated product model

The integrated product model is a further concept, the implementation of which enables a broad coupling of the different applications in the CIM area (cf. Chapter 3).

In contrast to a functional perspective (production model) on which the coupling concepts described above are based, now the coupling of applications is viewed from an object oriented point of view, i.e. with regard to the product as an object.

The main purpose of the product model is to create a common semantic structure covering all software systems which participate in the product oriented information and data flow, concerning the entire life cycle of a product.

The product model contains all product data of various sorts (geometrical, technological, organizational, etc.). The required resources which must not be regarded as separated from the product itself are also included in the product model and considered as products in their own right (PDES 1990).

The storage of information in the product model may follow the layer concept (for the following cf. Spur 1986, Krause 1986, Anderl 1989b).

For instance, information describing the shape of an object is stored in the information layer geometry. Each information layer represents a semantic class within the entire product model information. Link layers form a connection between the different information layers and, within one information layer, between information having a similar semantic structure. The link layer concept supports the consistency and the low redundancy of the overall model. To give an example, in case of an alteration of the geometry, the link layer between the geometry layer and NC programming layer automatically effects a corresponding alteration of the the NC program. A higher level organization layer provides the access mechanisms to the product model.

In spite of being a logical unit, the product model may be physically distributed.

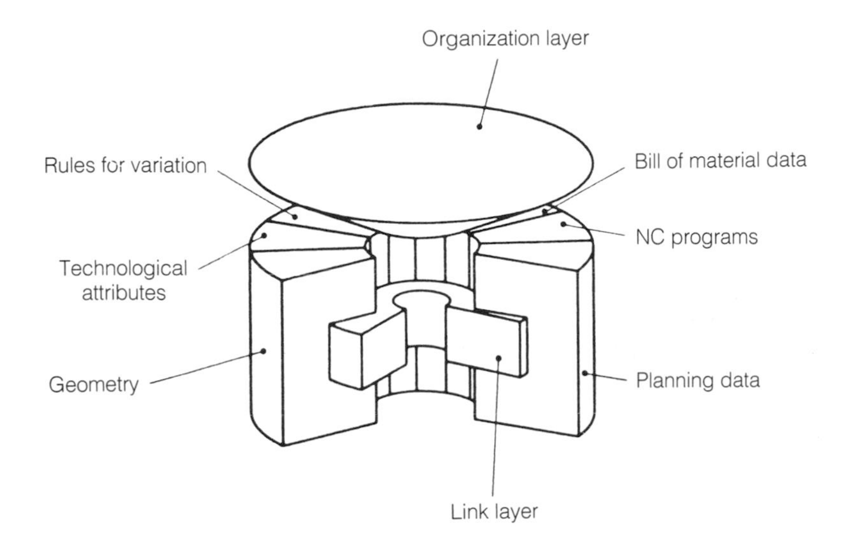

Fig. 7.6 Product model (Spur 1986).

The information layer concept has to some extent already been introduced into some CAD/CAM systems, although comprehensive systems have not yet been implemented. For achieving an integration via the product model, existing applications have to be adapted to a considerable extent, as in their current form they are running on an autonomous subset of the overall product model largely depending on the system used.

The product model concept will for some time to come remain the object of research and development.

7.6 CIM-OSA

In the framework of the project No. 688 CIM-OSA (Computer Integrated Manufacturing - Open System Architecture) within the ESPRIT Programme sponsored by the European Community, an open system architecture for CIM systems is currently being developed (CIMOSA 1989). Considering the information technological modules on the implementation level, the 'Integrating Infrastructure' was designed as the basis for a company-wide infrastructure.

It provides services for the management of business processes (process management), information management, data transfer (system-wide exchange), and interfaces to the outside world (front end services). It separates the application programs from system components as well as from the model specification (implementation model - business events) at well-defined interfaces.

The services are on the one hand divided into management and interface services, and on the other hand they are differentiated according to their relationships with functions, data, and the communication tasks.

The management services ensure the integration of events within the enterprise and enforce a system behaviour in accordance with the specifications of the implementation model, while the interface services integrate the functional elements and thus the applications.

The services for the management of business events can be subdivided into further categories. The business control service supervises and controls the business events lodged with the function view of the implementation model. The activity control performs the same task for the lodged functions. According to specifications by the resource view of the implementation model, the resource management service links the resources with the business events which are just being executed. The management services communicate with the functional elements only via the front end services for humans, machines, and programs. The three functional elements also communicate only via the corresponding front end services. This leads, for example, to a complete disconnection of application programs (functional element: program) and user interfaces (functional element: man).

According to specifications by the information view of the implementation model, the data management service handles all data, the place where they are stored, the schemata and the rights of access, and ensures their consistency. It is the only service to communicate with the front end service for data and the functional elements for data storage.

Its counterpart is the communication management service which handles the entire communication between humans, machines, or among services, as well as between services and the other elements. Here the front end service which is exclusively used follows the ISO/OSI reference model. This approach is currently still in the conceptualization phase (cf. Scholz-Reiter 1990).

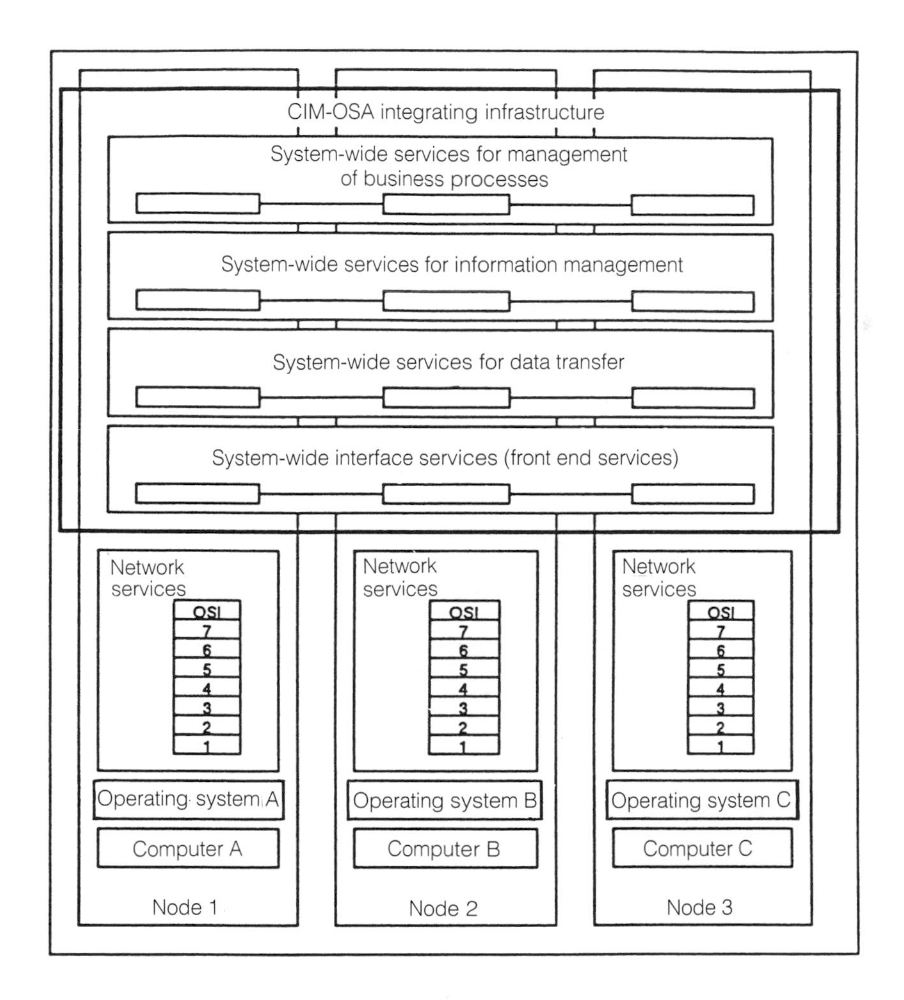

Fig. 7.7 System-wide services of the integrating infrastructure (CIMOSA 1989).

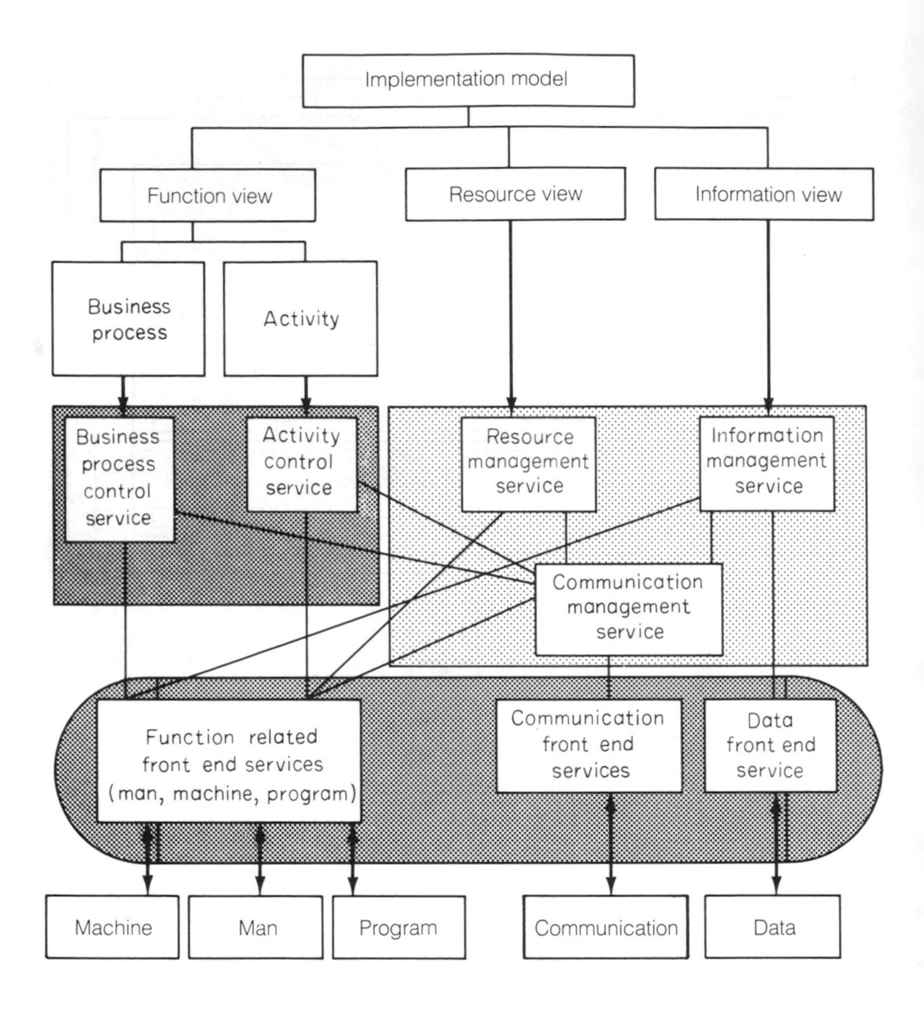

Fig. 7.8 Coupling of the services of the integrating infrastructure (CIMOSA 1989).

7.7 Summary

There are two basic strategies for putting into practice the theoretical requirements for a complete and functioning integration:

- The implementation of new CIM systems can be attempted which in their overall functional range are adjusted in such a way that the components they include fulfil the aims of data and function integration.
- However, the implementation of CIM can also be based on existing operational program packages. Then the existing functions with additional components are coupled, usually by means of an information and communication management system.

The latter strategy will for some time to come be of more relevance in actual practice.

Comprehensive concepts for coupling should ensure a factory or company wide administration which co-ordinates and couples all control and data flows.

The common characteristics of the presented concepts and their possible implementations can be summarized by the following statements:

- existing procedures can be integrated;
- lines of communication are shortened;
- the topicality and consistency of data is increased;
- the redundancy of data and operations is reduced;
- high expenditure for state management (starvation and deadlock problems);
- job control is likely to consume more of the overall system capacity than functional applications; and
- highly sophisticated algorithms or expert systems have to be applied.

References

Abel, E. and Heckl, H (1987). EDIF - Eine Initiative zur Standardisierung von Entwurfsdaten im Mikroelektronik-Bereich. In CAD-Schittstellen und Datenaustausch im Elektronik-Bereich (ed. M.H. Ungerer).Berlin.

AFNOR (1985). Automatisation Industrielle Representation externe des donnees de definition de produits. Specification du standard d'echange et de transfert (SET). Version 85-08.Z68-300. Association française de normalisation. Paris.

Albus, J.S., Barbera, A.J. and Nagel, R.N.(1981). Theory and Practise of Hierarchical Control. 23rd IEEE Computer Society International Conference, Compcon Falls, 18-39.

Anderl, R. (1986a). Schnittstellen zur Integration von CAD- und NC-Systemen. In Kommtech 86, Kongreß IV, Velbert, 40-1-40-22.

Anderl, R. (1986b). Schnittstellen zum Austausch produktdefinierender Daten von CAD-Systemen. Referat zum IBM-Kongreß 1986, Technisch-Wissenschaftliche Informationsverarbeitung in Hochschulen, Forschung und Industrie. Düsseldorf.

Anderl, R. (1987). Schnittstellen zum Austausch produktdefinierender Daten. FB/IE 36(1987)4, 160-166.

Anderl, R. (1989a). Normung von CAD-Schnittstellen. CIM Management 5(1989)1, 4-8.

Anderl, R. (1989b). Integriertes Produktmodel. ZWF/CIM 84(1989)1, 640-644.

ANSI (1981). Digital Representation of Product Definition Data. ANSI Standard Y14.26M.

ANSI (1986). Graphical Kernel System for three dimensions (GKS-3D). Working document, American National Standards Institute, Technical Committee X3H3 - Computer Graphics, Accredited Standards Committee X3. Information Processing Systems, Irvine, CA.

ANSI/ISO (1986a). Database Language SQL Document ANSI X3.135-1986, Document ISO/TC97/SC21/WG3 N 117.

ANSI/ISO (1986b). Database Language SQL Addendum-2 (working draft). Document X3.H2-86-61 (May 1986), Document ISO/TC97/SC21/WG3 N 143.

Appleton, D.S. (1986). Introducing the New CASA CIM Wheel. Computer And Automated Systems Association of SME. Dearborn, MI.

Armitage, B., Dunlop, G., Hutchinson, D. and Yu, S. (1988). Fieldbus: An emerging communications standard. Microprocessors and Microsystems, 12, No. 10, 555-562.

Aschenbrenner, J.R. (1986). Open System Interconnection. IBM Systems Journal. 25, Nos. 3/4, 369-379.

ATP (1988). Offene Kommunikation jetzt auch im Feldbusbereich. atp 30(1988)1, 50.

AWF (Ausschuß für Wirtschaftliche Fertigung e.V.) (ed.) (1985). AWF-Empfehlung - Integrierter EDV-Einsatz in der Produktion - CIM Computer Integrated Manufacturing - Begriffe, Definition Funktionszuordnungen. Eschborn.

Bauer, M. (ed.) (1987). Automatisierungssystem MAP. Berlin.

BERKOM (1987a). Protokolle der BERKOM Arbeitskreise Detecon, Potsdamer Str. 87, 1000 Berlin 30.

BERKOM (1987b). BERKOM Architekturmodell Detecon, Potsdamer Str. 87, 1000 Berlin 30.

Bey, I. and Leuridan, J. (1986). Europäische Vorhaben zur Definition von CAD-Schnittstellen. ZWF 81(1986)1, 38-42.

Bey, I. and Gengenbach, U. (1988). CAD*I interface for solid model exchange. Computer Graphics, 12, No. 2, 181-190.

Bey, I. and Stanek, J. (1989). Schnittstellen zwischen CAD-Systemen. CAD-CAM Report (1989)4, 48-61.

Blacher, A., Dabrowski, R. and Scholz-Reiter, B. (1990). Analysis and design of information and communication structures in CIM. Computer Integrated Manufacturing Systems, 3, No. 3, 135-140.

Borst, W., Lindner, K.-P. and Ziesemer, M. (1988). Der EUREKA-Feldbus für die Instrumentierungstechnik der 90er Jahre. atp 30(1988)9, 430-435.

Braendli, N. and Mittelstaedt, M. (1989). Exchange of solid models, current state and future trends. Computer Aided Design, 21, No. 2, 87-96.

Brenig, H. (1990). Informationsflußbezogene Schnittstellen bei industriellen Produktionsprozessen. Information Management 5(1990)1, 28-39.

Bullinger, H.J., Lay, K. and Warschat, J. (1987). Perspektiven aus der Integration von CA-Komponenten. Technische Rundschau, 22, 54-60.

CAM-I (1983). Conceptional information model of an advanced factory management system - work center level. Final Report R-83-FM-01;CAM-I, Arlington, Texas.

Caruso, M. (1988). The VISION Object Oriented Database Management System Proceedings of OOPSLA '88, San Diego, CA.

Chorafas, D.N. (1987). Engineering productivity through CAD/CAM. Butterworth, London.

Chorafas, D.N. (1989). Handbook of Database Management and Distributed Relational Databases. TAB, Blue Ridge Summit, PA.

Chryssolouris, G. and Gruenig, I. (1988). On a database design for intelligent manufacturing systems. International Journal of Computer Integrated Manufacturing, 1, No. 3, 171-184.

CIM (1986). Was ist heute Stand der Technik bei Netzwerk-Herstellern. CIM Management 2(1986)3, 34-49.

CIMOSA (1989). CIM-OSA Reference Architecture Specification. ESPRIT Project No. 686, Brussels.

Codd, E.F.(1990). The Relational Model for Database Management - Version 2. Addison-Wesley, Reading, MA.

Cooling, J. and Hussein, S. (1989). Shortcomings of Mini-MAP in high speed real time local area networks. Proc. Second International Conference on Software Engineering for Real Time Systems, IEE Publication No. 309, Stevenage, 194-198.

Craig, I.D. (1989). The Cassandra architecture - distributed control in a blackboard system. Chichester, UK.

Dadam, P. *et al.* (1986). Design of an integrated DBM to support advanced applications. Informatik Fachberichte 94. Berlin, 363-381.

Dadam, P. and Linnemann, V. (1989). Advanced Information Management (AIM): Advanced database technology for integrated applications. IBM Systems Journal, 28, No. 4, 661-681.

Danner, W.F. (1990). A Proposed Framework for Product Data Modeling. U.S. Department of Commerce, NIST, Gaithersburg, MD.

Date, G.J. (1987). An introduction to database systems, (4th edn). Addison-Wesley, Reading, MA.

Davies, R. and Smith, R.G. (1983). Negotiation as a Metaphor for Distributed Problem Solving. Artificial Intelligence, 14(1983), 63 ff.

Deppisch, U. *et al.* (1986). Überlegungen zur Datenbank-Konzeption zwischen Server und Workstations. Proc. 16. GI Jahrestagung, Berlin, 565-580.

DIN (1974). DIN 66215: CLDATA.

DIN (1983). DIN 66025: Programmaufbau für numerisch gesteuerte Arbeitsmaschinen.

DIN (ed.) (1987a). Format zum Austausch von Normteildateien, DIN V 66304.

DIN (ed.) (1987b). CAD-Normteiledatai nach DIN, DIN-Fachbericht 14; Berlin, Köln.

DIN (ed.) (1987c). Normung von Schnittstellen für die rechnerintegrierte Produktion, DIN-Fachbericht 15; Berlin, Köln.

DIN (ed.) (1989a). Schnittstellen der rechnerintegrierten Produktion (CIM) - CAD und NC-Verfahrenskette, DIN-Fachbericht 20; Berlin, Köln.

DIN (ed.) (1989b). Schnittstellen der rechnerintegrierten Produktion (CIM) - Fertigungssteuerung und Auftragsabwicklung, DIN-Fachbericht 21; Berlin, Köln.

Dittrich, K.R. (1988). Advances in Object-Oriented Database Systems. Proc. of the 2nd Int. Workshop on Object-Oriented Database Systems, Lecture Notes in Computer Science 334. Springer, Berlin.

Duelen, G., Linnemann, H. and Bernhardt, R. (1986). Die Informationsarchitektur in datengetriebenen Fabriken. Vorträge PTK 86, ZWF/CIM 81(1986)11, 56-66.

Eberlin, W. (1984). CAD-Datenbanksysteme. Berlin.

Eckhardt, K.J. and Nowack, R. (1988). Standard-Architekturen für Rechnerkommunikation. Handbuch der Informatik 7.1. München, Wien.

Effelsberg, W. (1987). Datenbankzugriff in Rechnernetzen. it29(1987)3, 140-152.

Eigner, M. and Maier, H. (1985). Einstieg in CAD. München, Wien.

Eigner, M., Rüdigner, W. and Schmich, M. (1986). Kopplung von CAD mit PPS- und Informationssystemen als Baustein eines CIM-Konzepts. ZWF/CIM 81(1986)11, 611-614.

Elmasri, R. and Navathe, S.B. (1989). Fundamentals of database systems. Benjamin/Cummings, Redwood City, CA.

Encarnacao, J.L., Encarnacao, L.M. and Herzner, W. (1987). Graphische Datenverarbeitung mit GKS. München, Wien.

Enderle, G., Kansy, K. and Pfaff, G. (1987). Computer graphics programming: GKS, the graphic standard, (2nd rev. edn). Springer, Berlin.

Enderle, G. and Scheller, A. (ed.) (1989). Normen der graphischen Datenverarbeitung. Handbuch der Informatik 9.1. München, Wien.

Engel, H.O. (1990). FELDBUS-Normung 1990. atp 32(1990)6, 271-277.

Ernst, G. and Reubsaet, G. (1984). CADCPL - Weiterverwendung von Modelldaten aus CAD-Systemen. Industrie Anzeiger 106(1984)22, 34-41.

ESP (1984). IGES Experimental Solid Proposal. Internal IGES Paper.

Feiten, L. (1990). Lokale Netze. In Lexikon der Wirtschaftsinformatik, 2. vollst. neubearbeitete und erweiterte Auflage (ed. P. Mertens). Berlin, 261-263.

Felts, W.J. (1988). 1990 ISDN - the plan to achieve nationwide compatibility. Proceedings of IEEE Global Telecommunications Conference & Exhibition: Communications for the Information Age, IEEE, Piscataway, NJ, 924-927.

Fischer, W.E. (1982). Das Datenbanksystem PHIDAS als Werkzeug für Entwurf, Realisierung und Integration von Produktmodellen. Informatik Fachberichte 65, 107-119.

Förster, H.U. and Syska, A. (1985). Rechnerintegrierte Produktion - Ergebnisse einer Umfrage zu Verbreitung und Entwicklungstendenzen von EDV-Systemen in der Produktion. FIR-Mitteilungen, Sonderdruck 2/85, Aachen.

Fowler, J.E. (1989). System for creating and manipulating mechanical part models based on PDES. Proceedings of Second International Conference on Data and Knowledge Systems for Manufacturing and Engineering, Piscataway, NJ, 142-143.

Fowler, J.E. (1990). Development Plan STEP Production CELL; NISTIR 4421, National Institute of Standards and Technology.

Fox, M.S. and Smith, S.F. (1984). ISIS - a Knowledge-Based System for Factory Scheduling. Expert Systems, the International Jounal of Knowlege Engineering, Vol. 1, Learned Information Inc., Medford, N.J.

Geitner, U.W. (1991). Die CIM-Konzeption. In CIM-Handbuch (2. vollständig überarbeitete und erweiterte Auflage), Braunschweig, 3-4.

Göbel, K.-J., Hidde, A. and Schwartz, R. (1990). Concept definition of a logistics standard interface for CIM implementations. International Journal of Computer Integrated Manufacturing, 3, No. 5, 289-298.

Grabowski, H. and Glatz, R. (1986). Schnittstellen zum Austausch produktdefinierender Daten. VDI-Z 128(1986)10, 333-343.

Grabowski, H. and Glatz, R. (1987). IGES Model Comparison System: A Tool for Testing and Validating IGES. IEEE Computer Graphics and Applications, 7, No. 11, 47-57.

Grabowski, H., Anderl, R., Schilli, B. and Schmitt, M. (1989). STEP - Entwicklung einer Schnittstelle zum Produktdatenaustausch. VDI-Z 131(1989)9, 68-76.

Gröner, L. and Roth, L. (1987). CIM-Handler für die Verbindung von Softwaresystemen. CIM Management 3(1987)4, 14-19.

Graves, R.G., Yelamanchili, B. and Parks, C.M. (1988). An interface architecture for CAD/CAPP integration using knowledge-based systems and feature recognition algorithms. International Jounal of Computer Integrated Manufacturing, 1, No. 2, 89-100.

Groover, M.P. and Zimmers Jr., E.W. (1984). CAD/CAM Computer Aided Design and Manufacturing. Prentice-Hall, Englewood Cliffs, NJ.

Hackstein, R. and Braun, M. (1990). CAD und PPS koppeln - eine komplexe Gestaltungsaufgabe. CIM Management 6(1990)3, 64-67.

Härder, T., Meyer-Wegener, K., Mitschang, B. and Sikeler, A. (1987). PRIMA - A DBMS Prototype Supporting Engineering Applications. Proceedings 13th VLDB Conference, Brighton, U.K., 433-442.

Harding, B. (1988). EDA vendors cooperate on EDIF and proposed CAD framework standard. Computer Design, 27, No. 15, 31-32, 35.

Harmon, P. and King, D. (1985). Expert Systems: Artificial Intelligence In Business. Wiley, New York.

Harhalakis, G., Mark, L., Bohse, M. and Cochrane, B. (1987). Integration of Manufacturing Resource Planning (MRP II) and Computer Aided Design (CAD) based on Update Dependencies. Proceedings International Conference on Data and Knowledge Systems for Manufacturing and Engineering, IEEE, NY, 83-92.

Harrington, J. (1973). Computer Integrated Manufacturing. New York.

Harrington, J. (1984). Understanding the Manufacturing Process. New York.

Hayes-Roth, B. (1985). A Blackboard Architecture for Control. Artificial Intelligence 26(1985), 251-321.

Heilmann, H. (1989). Integration: Ein zentraler Begriff der Wirtschaftsinformatik im Wandel der Zeit. HMD 150/1989, 46-58.

Helberg, P. and Wunderlich, L. (1985). Computerintegrierte Fertigung (CIM) - Anforderungen an die Informationsverarbeitung - Eine Studie in der deutschen Fertigungs- und Grundstoffindustrie. Arbeitspapier TU, Berlin.

Helberg, P. (1987). PPS als CIM-Baustein. Berlin.

Hellwig, H.-E., Hellwig, U. and Paulus, M. (1983a). Die Kopplung von CAD und CAM, Teil 1: Mögliche Schnittstellen und ihre Vor- und Nachteile. VDI-Z 125(1983)10, 355-360.

Hellwig, H.-E., Hellwig, U. and Paulus, M. (1983b). Die Kopplung von CAD und CAM,. Teil 2: Der Informationsfluß von der Konstruktion zur Fertigung. VDI-Z 125(1983)11, 454-460.

Hellwig, H.-E., Hellwig, U. and Paulus, M. (1985). Die Kopplung und die Integration von CAD und CAM, Teil 3: CAD/NC-Kopplung. VDI-Z 127(1985)1/2, 28-32.

Hellwig, H.-E. and Hellwig, U. (1987). Schnittstellen. In CIM Handbuch (ed. U.W. Geitner), Braunschweig, 202-208.

Henn, O. (1990). PROFIBUS. CIM Management 6(1990)4, 55-62.

Hirsch-Kreinsen, H. (1986). Technische Entwicklungslinien und ihre Konsequenzen für die Arbeitsgestaltung. In Rechnerintegrierte Produktion - Zur

Entwicklung von Technik und Arbeit in der Metallindustrie (ed. H. Hirsch-Kreinsen and R. Schultz-Wildt), München.

Hofmann, J. (1988). Aktionsorientierte Datenverarbeitung im Fertigungsbereich, Berlin.

Hübel, Ch., Sutter, B. and Paul, R. (1990). Datenbankgestützte technische Modellierung - ein Ansatz für die CAD/CAP-Integration. CIM Management 6(1990)2, 48-54.

Hüllenkremer, M. and Viehmann, K. (1986). CAD/CAP - Integration. CAE-Journal 5/86, 44-47.

Integrated Computer-Aided Manufacturing (ICAM) (1978). Task 1 - Final Report Manufacturing Architecture, AFML-TR-78-148. Wright-Patterson Air Force Base, Ohio.

Integrated Computer-Aided Manufacturing (ICAM) (1981). Architecture Part II: Composite Function Model of 'Manufacturing Product' (MFG0). Wright-Patterson Air Force Base, Ohio.

Initial Graphics Exchange Specification (IGES) (1983). Version 2.0. National Bureau of Standards.

Initial Graphics Exchange Specification (IGES) (1986). Version 3.0. National Bureau of Standards.

Initial Graphics Exchange Specification (IGES) (1988). Version 4.0. National Bureau of Standards.

ISO (1985). External Representation of Product Definition Data. A Status Report of ISO Subcommittee TC 184/SC4, ISO Internal paper.

ISO (1990). Exchange of Product Model Data - Part 11: The EXPRESS Language. ISO CD 10303 - 11, TC 184/SC4N64/WG1, ISO Internal paper.

Ito, Y., Shinno, H. and Saito, H. (1990). Proposal for CAD/CAM interface with expert systems. Robotics and Computer Integrated Manufacturing, 4, Nos. 3/4, 491-497.

Jones, A., Barkmeyer, E. and Davis, W. (1989). Issues in the design and implementation of a system architecture for computer integrated manufacturing. International Journal of Computer Integrated Manufacturing, 2, No. 2, 65-76.

Jones, A. and Saleh, A. (1990). A multi-level/multi-layer architecture for intelligent shopfloor control. International Journal of Computer Integrated Manufacturing, 3, No. 1, 60-70.

Kalashian, M.A. (1990). EDI - A critical link in customer responsiveness. Manufacturing Systems, 8, No. 12, 20-23.

Katz, M., Biwer, G. and Bender, K. (1989). Die PROFIBUS-Anwendungsschicht. atp 31(1989)12, 588-597.

Kauffels, H.J. (1985). Lokale Netze. Köln.

Kim, W. (1990). Introduction to Object-Oriented Databases. MIT Press, Cambridge, MA.

Klement, K. and Nowacki, H. (1988). Exchange of Model Presentation Information between CAD Systems. Computers & Graphics, 12, No. 2, 173-180.

Knappe, H.J. and Veerkamp, H.J. (1986). Integration der NC-Programmierung. Industrie Anzeiger 108(1986)20, 35-38.

Köhl, E., Esser, U., Kemmner, A. and Wendering, A. (1988). Auswertung der CIM-Expertenbefragung. FIR, Aachen.

Köhl, E., Esser, U., Kemmner, A. and Förster, U. (1989). CIM zwischen Anspruch und Wirklichkeit - Erfahrungen, Trends, Perspektiven. Eschborn, Köln.

Kölling, J. (1989). CAD-Normteiledatei nach DIN. CIM Management 5(1989)1, 30-36.

Kommtech (1986). 5th event, several contributions. Velbert.

Krallmann, H. (1986). CIM zur Verbesserung der internationalen Wettbewerbsfähigkeit. In Innovation und Wettbewerbsfähigkeit (ed. E. Dichtl, W. Gerke and A. Kieser). Wiesbaden, 197-225.

Krause, F.-L. (1986). Fortgeschrittene Konstruktionstechnik durch neue Softwarestrukturen. Presentations to the PTK 86, ZWF/CIM81(1986)11, 114-123.

Küspert, K. (1986). Das Aktuelle Schlagwort: Non-Standard-Datenbanksysteme. Informatik Spektrum 9(1986)3, 184-195.

Latz, H.W. (1986). Welche CAD-Systeme sind CIM-tauglich. CIM Management 2(1986)1, 13-17

Lorie, R.A. *et al.* (1985). Supporting Complex Objects in a Relational System for Engineering Databases. Query Processing in Database Systems, Berlin, 145-155.

Lukasik, V.J. (1986). TOP - Technical and Office Protocols. Proceedings of the National Electronics Conference, 40, Pt 1, Professional Education Int. Inc., 229-247.

van Maanen, J. and Leuridan, J. (1990). Methods for exchange of finite element analysis data and integration of finite element analysis and prototype testing. International Journal of Computer Integrated Manufacturing, 3, Nos. 3 and 4, 181-185.

Madron, T.W. (1988). Local Area Networks - The second Generation. Wiley, New York.

Magill, W.R. and McLeod, A.J. (1989). Automated Generation of NC part programs from a feature-based component description. International Journal of Computer Integrated Manufacturing, 2, No. 4, 194-204.

Maier, H. (1986). Datentechnische Möglichkeiten und Probleme der CAD/CAM-Integration. In Rechnerintegrierte Produktion - Zur Entwicklung von Technik und Arbeit in der Metallindustrie (ed. H. Hirsch-Kreinsen and R. Schultz-Wild). München, 49-80.

Martin, J. (1983). Managing the Database Environment. Prentice-Hall, Englewood Cliffs, NJ.

Martin, J. and Chapman, K.K. (1989). Local Area Networks - Architectures and Implementations. Prentice-Hall, Englewood Cliffs, NJ.

McLean, Ch., Mitchell, M. and Barkmeyer, E. (1983). A computer architecture for small-batch manufacturing. IEEE spectrum, May 1983, 59-64.

Meen, S., Oian, J. and Ulfsby, S. (1982). TORNADO - a DBMS for CAD/CAM Systems. Proceedings of the IFIP WG 5.2 Working Conference, North Holland, 335-346.

Meijer, A.H. (1989). Electronic data interchange for product structure data. International Journal of Computer Integrated Manufacturing, 2, No. 4, 220-228.

Mertens, P. and Hofmann, J. (1986). Aktionsorientierte Datenverarbeitung. Informatik Spektrum 9(1986)6, 323-333.

Messer, B. (1988). Breitband-ISDN als Kommunikationsmedium für CIM. CIM Management 4(1988)3, 61-66.

Milberg, J. and Peiker, St. (1987). Geometrie- und technologieorientierte Verbindung von CAD-Systemen mit NC-Programmiersystemen. wt77(1987),583-586.

Mills, R. (1989). 15 ways to link CAD and CAM. CAE - Computer Aided Engineering, 8, No. 10, 5.

Mouleeswaran, C.B. and Fischer, H.G. (1986). A Knowledge-Based Environment for Process Planning. Siemens Corporate Research and Support, Inc., Princeton, NJ.

Mufti, A.A., Morris, M.L. and Spencer, W.B. (1990). Data exchange standards for computer-aided engineering and manufacturing. International Journal of Computer Applications in Technology, 3, No. 2, 70-80.

Mund, A. *et al.* (1987). VDA-Flächenschnittstelle (VDAFS) Version 2.0. VDA-Arbeitskreis CAD/CAM Verband der Automobilindustrie e.V. (VDA), Frankfurt.

Nakagawa, K., Matsudaira, T. and Shiobara, Y. (1988). Distributed control system utilizing a MAP-based LAN. Advances in Instrumentation, Proceedings, 43, Pt 2, ISA, Research Triangle Park, NC, 559-572.

Nakano, N. and Mizuno, T. (1990). MAP-compatible equipment for FA networks. Mitsubishi Electric Advance, 53, 16-19.

Nedeß, C. and Landvogt, F.-B. (1986). Rechnerintegrierte Auftragsabwicklung. VDI-Z 12(1986)14, 540-546.

Nii, P.H. (1986). Blackboard Systems: The Blackboard Model of Problem Solving and the Evolution of Blackboard Architecures. The AI Magazine, Summer 1986, 38-53.

Nittel, S. (1989). Relationale und objektorientierte Datenbanksysteme für CIM-Applikationen - ein Vergleich. CIM Management 5(1989)6, 11-14.

Noll, S., Poller, J. and Rix, J. (1987). Approach to solving the compatibility problem between GKS and PHIGS. Computer Graphics, 19, No. 8, 456-463.

Norsk Data (ed.) (1986). Integrierte Datenverarbeitung für Konstruktion und Arbeitsplanung. München, Wien.

Nowacki, H. (1990). Austausch von CAD-Daten über Rechnernetze. CIM-Management 6(1990)2, 4-9.

Osland, C.D. (1988). GKS and CGM graphics standards. Computer Physics Communications, 50, No. 1/2, 129-141.

Paul, H.P., Schek, H.J., Scholl, M.H., Weikum, G. and Deppisch, U. (1987). Architecture and implementation of the Darmstadt database kernel system. Proc. ACM SIGMOD 1987 Annual Conference on Management of Data, 196-207.

PDES (1985). The content, plan, and schedule for the first version of the Product Data Exchange Specification (PDES). Internal IGES paper.

PDES (1990a). Product Life Cycle Support (PLCS) Documentation Package, Draft. NIST PLCS Committee.

PDES (1990b). IGES/PDES organization - reference manual. National Computer Graphics Association, Fairfax, Virginia.

Pfeifer, T. and Heiler, K.-U. (1987). Ziele und Anwendungen von Feldbussystemen. atp 29(1987)12, 549-557.

Pham, T.T. (1982). Erfahrungen mit CAD und der NC-Koppelung. ZWF (1982)10, 453-464.

Pimentel, J.R. (1990). Communication Networks for Manufacturing. Englewood Cliffs, NJ.

Rake, H., Kaster, L. and Zebermann, Ch. (1989). PROFIBUS in der nicht eigensicheren Verfahrenstechnik. CIM Management 5(1989)5, 15-20.

Ranky, P.G. (1986). Computer Integrated Manufacturing. Prentice-Hall, Englewood Cliffs, NJ.

Roach, J.M. (1990). MAP 1995. ISA Transactions, 29, No. 1, 47-51.

Scheer, A.-W. (1988). CIM Computer Steered Industry. Springer, Berlin.

Scheer, A.-W., Herterich, R. and Klein, J. (1990). INMAS - Eine individuell konfigurierbare Schnittstelle. Information Management 5(1990)1, 16-26.

Scheuernstuhl, G., Schneider, H.J. and Wild, J.-K. (without date). Manuskript zur CIM-Lehrveranstaltung Datenbanksysteme I. Fachbereich Informatik, TU Berlin.

Schlechtendahl, E.G. (ed.) (1988). Specification of CAD*I Neutral File for CAD Geometry. Berlin.

Schlechtendahl, E.G. (ed.) (1989a). CAD Data Transfer for Solid Models. Berlin.

Schlechtendahl, E.G. (ed.) (1989b). The interaction of CAD*I with ISO/TC184/SC4/WG1. In CAD Data Transfer for Solid Models (ed. E.G. Schlechtendahl). Berlin.

Schmidt, P. (1986). CAD-Datenaustausch aus der Sicht eines mittelständischen Automobilzulieferers. Proceedings CAT 1986, Stuttgart, 38-41.

Scholz, B. (1985). Zukunft mit CIM. CIM Management 1(1985)4, 11-18.

Scholz, B (1987). Das CIM-Modell der CASA/SME (USA). CIM Management 3(1987)2, 54-55.

Scholz, B. *et al.* (1987a). CIM im Werkzeugbau - Istanalyse. Projektbericht Systemanalyse II, TU Berlin.

Scholz, B. *et al.* (1987b). CIM im Werkzeugbau - Sollkonzept. Projektbericht Systemanalyse II, TU Berlin.

Scholz, B. and Neidhardt, J. (1988). Eine Realisierung der Stücklistenübergabe zwischen dem CAD-System CADIS-2D und dem PPS-System ISI. CIM-Management 4(1988)3, 53-60.

Scholz-Reiter, B. (1990). CIM-Informations- und Kommunikationssysteme - Darstellung von Metoden und Konzeption eines rechnergestützten Werkzeugs für die Planung. München, Wien.

Scholz-Reiter, B. (1991). Schnittstellen für die rechnerintegrierte Produktion; In CIM-Handbuch (2. vollständig überarbeitete und erweiterte Auflage) (ed. U.W. Geitner), Braunschweig, 605-624.

Schümmer, M. (1988). Manufacturing Message Specification MMS/RS-511. Informatik Spektrum 11(1988)4, 209-211.

Schultz-Wild, R., Nuber, Ch., Rehberg, F. and Schmierl, K. (1989). An der Schwelle zu CIM - Strategien, Verbreitung, Auswirkungen. Eschborn, Köln.

Schwarz, K. (1989). Manufacturing Message Specfcation (MMS). atp 31(1989)1, 23-29.

Schwindt, P. (1986). CAD-Datenaustausch aus der Sicht eines mittelständischen Automobilzulieferers. In CAT 86-Proceedings, Stuttgart, 38-41.

Seifert, H. (1986). Von CAD zu CIM. VDI-Z 128(1986)10, 327-331.

Sedlaczek, D. (1988). Kommunikationsstrukturen in CIM. In Unterlagen zum Seminar Computer Integrated Manufacturing (ed. B. Scholz), TU Berlin.

SET (1984). SET Standard d'Echange et de Transfert Specification Rev. 1.1.. Aerospatiale.

Shioe, Y. and Lukehart, W. (1988). 10-Mbps carrier band token bus for process control supervisory network. Advances in Instrumentation, Proceedings, 43, Pt 2, ISA, Research Triangle Park, NC, 547-558.

Smith, S.F., Fox, M.S. and Ow, P.S. (1986). Constructing and Maintaining detailed Production Plans: Investigations into the Development of Knowlege-Based Factory Scheduling Systems. AI Magazine.

Spur, G. and Krause, F.-L. (1984). CAD-Technik. München, Wien.

Spur, G. (1986). CIM - Die informationstechnische Herausforderung an die Produktionstechnik. Vorträge PTK 86, ZWF/CIM 81(1986)11, 5-19.

Ssemakula, M.E. and Satsangi, A. (1989a). Application of PDES to CAD/CAPP integration. Computers & Industrial Engineering, 18, No. 1-4, 234-239.

Ssemakula, M.E. and Rangachar, R.M. (1989b). Prospects of process sequence optimization in CAPP systems. Computers & Industrial Engineering, 16, No. 1, 161-170.

Ssemakula, M.E. (1990). Process planning system in the CIM environment. Computers & Industrial Engineering, 19, No. 1-4, 452-456.

Stallings, W. (1989). ISDN - an introduction. Macmillan, New York.

Steinmetz, G. (1991). Grunddatenverwaltung. In CIM-Handbuch. (2. vollständig überarbeitete und erweiterte Auflage) (ed. U.W. Geitner), Braunschweig, 54-67.

STEP (1984). International Standardization Organization (ISO): STEP (Standard for the Exchange of Product Model Data) Requirements Document, ISO TC184/SC4/WG1N 4.

STEP (1988). STEP Preliminary Design; ISO TC184/ SC4/WG1N 208.

STEP (1990a). Product Life Cycle Support (PLCS) Documentation Package; ISO TC184/SC4; ISO internal paper, draft version.

STEP (1990b). STEP Part 1: Overview and fundamental principles version 4; ISO TC184/SC4/WG6 N6.

Steusloff, H.U. (1987). Kommunikation in Produktionsunternehmen. In Automatisierungssystem MAP (ed. M. Bauer). Berlin.

Steusloff, H.U. (1989). Funktionsstruktur, Kommunikationsstruktur und Kommunikationsmittel in Automatisierungs- und Leitmitteln. atp 31(1989)5, 209-217.

Stonebraker, M. and Rowe, L.A. (1986). The Design of POSTGRES. Proceedings ACM SIGMOD Conference, Washington D.C., 340-355.

Suppan-Borowka, J. and Simon, T. (1987). MAP-Datenkommunikation in der automatisierten Fertigung. Pulheim.

Suppan-Borowka, J. (1987a). Herstellerübergreifend mit MAP. Computer Magazin 10, 53-55.

Suppan-Borowka, J. (1987b). TOP-Technical and Office Protocols. Informatik Spektrum 10(1987)4, 218-220.

Taylor, F.W. (1911). The Principles of Scientific Management. New York.

TUBKOM (1987). Projektvorschlag für ein multifunktionales Breitband-Kommunikationssystem. TU Berlin.

Ullmann, W. and Dräger, H. (1987). Organisation der Steuerung des Werkzeugbaus weist Schwachstellen auf. Industrie-Anzeiger 82, 42-43.

VDA/VDMA (ed.) (1983). VDA-Flächenschnittstelle (VDAFS) Version 1.0; Verband der Automobilindustrie.

VDMA/VDA (ed.) (1987). Festlegung einer Untermenge von IGES Version 3.0 (VDA-IS), VDMA/VDA 66319.

Wanhao, L. and Switzer, H. (1989). Unified data exchange based on EDIF. Proceedings - Design Automation Conference, IEEE Piscataway, NJ, 803-806.

Weatherall, A. (1988). Computer Integrated Manufacturing, Butterworth, London.

Wedekind, H. (1986). Integrierte Fertigungsdatenbanken. Tagungsband GI-Tagung 1986, Berlin, 90-107.

Wedekind, H. (1987). Fertigungsdatenbanken. Informatik Spektrum 10 (1987)10, 40-41.

Weiss, J.A. (1984). Product Definition Data Interface: The Solution. ICAM Project Material.

Weston, R.H., Gascoigne, J.D., Rui, A., Hodgson, A., Sumpter, C.M. and Coutts, I. (1988). Steps towards information integration in manufacturing. International Journal of Computer Integrated Manufacturing, 1, No. 3, 140-153.

Weston, R.H., Gascoigne, J.D., Rui, A., Hodgson, A., Sumpter, C.M. and Coutts, I. (1989). Configuration methods and tools for manufacturing systems integration. International Journal of Computer Integrated Manufacturing, 2, No. 2, 77-85.

Wolf, M. (1986). Aktuelle Entwicklungen im LAN-Bereich. CIM-Management 2(1986)3, 14-18.

Wood, G.G. (1988). Current fieldbus activities. Computer Communications, 11, No. 3, 118-123.

Yeomans, R.W., Choudry, A. and Ten Hagen, P.J.W. (ed.) (1985). Design Rules For A CIM System. Amsterdam.

XBF (1981). Geometric Modeling Project, Boundary File Design (XBF-2). Report No. R-81-GM-O2.1. CAM-I Inc., Arlington, Texas.

ZVEI (ed.) (1987). Lokale Netze - Kommunikation im industriellen Bereich. I + K Spektrum 11, Frankfurt.

Index

E

I, J

K, L

Q, R

U,V,W,X,Y,Z

C. Dennis Pegden